Mimetic Discretization Methods

Mimetic Discretization Methods

José E. Castillo
San Diego State University
California, USA

Guillermo F. Miranda
Universidad Central de Venezuela
Venezuela

CRC Press
Taylor & Francis Group
Boca Raton London New York

CRC Press is an imprint of the
Taylor & Francis Group, an **informa** business

A CHAPMAN & HALL BOOK

CRC Press
Taylor & Francis Group
6000 Broken Sound Parkway NW, Suite 300
Boca Raton, FL 33487-2742

First issued in paperback 2019

© 2013 by Taylor & Francis Group, LLC
CRC Press is an imprint of Taylor & Francis Group, an Informa business

No claim to original U.S. Government works

ISBN-13: 978-1-4665-1343-3 (hbk)
ISBN-13: 978-0-367-38043-4 (pbk)

Visit the Taylor & Francis Web site at
http://www.taylorandfrancis.com

and the CRC Press Web site at
http://www.crcpress.com

Contents

List of Figures

List of Tables

List of Algorithms

Symbol Description

$\triangleq$	Equals by definition.
Ω	Usually represents an arbitrary volume.
σ	Usually represents an arbitrary surface.
$\Phi_{we}(\sigma)$	Usually represents the flux through an arbitrary surface σ in the west-to-east direction.
p_h	Cell projection operator.
P_h	Nodal projection operator.
$< \cdot, \cdot >_Q$	Weighted inner product with weight matrix Q.
div	Continuous differential operator: divergence.
$grad$	Continuous differential operator: gradient.
$curl$	Continuous differential operator: curl (or rotor).
DIV	Discrete analog to the continuous divergence operator.
GRAD	Discrete analog to the continuous gradient operator.
CURL	Discrete analog to the continuous curl operator.
$\check{D}$	Discrete Castillo–Grone divergence operator.
$\check{G}$	Discrete Castillo–Grone gradient operator.
$\check{L}$	Discrete Castillo–Grone Laplacian operator.
$\check{C}$	Discrete Castillo–Grone curl operator.
$\check{B}$	Discrete Castillo–Grone boundary operator.

Preface

Loosely speaking, "mimetic" or "compatible" algebraic discretization methods employ discrete constructs in an effort to mimic the continuous identities and theorems found in vector calculus. Many of the discretization methods provided in the literature attempt to achieve this compatibility via a variety of different approaches; the degree of success varies among them, both in terms of their order of accuracy and the variety of problems that they are able to solve.

Finite difference, finite element, and finite volume are all methods capable of transforming continuous models and equations into their discrete counterparts. Although the introduction of these discretization techniques has often been motivated by engineering and physical problems, unfortunately, they have not been able to produce compatible discrete models that fully mimic the continuous phenomenon being studied. However, among the various compatible discretization methods that have recently been developed, there are two that are particularly noteworthy. The first is the "compatible method" presented by P. Bochev, and the second is the mimetic discretization based on J. E. Castillo and R. H. Grone's mimetic differential operators. (The latter is the method on which this book is based.)

Bochev's methods feature an algebraic topology-based framework for mimetic discretization which utilizes a combinatorial version of Stokes' theorem and supports mutually consistent operations of differentiation and integration. This framework, which is both more general and more abstract, provides insight into many interesting questions related to compatible discretization techniques. However, its treatment of boundary value problems via Sobolev spaces is somewhat restrictive, particularly in regards to the flux throughout the boundary, as zero boundary values are not attained in a classical, continuous manner.

The mimetic discretization method from the Castillo–Grone operators is exclusively based upon an extended form of Gauss' divergence theorem, rather than on the more general Stokes' theorem. This method is simpler, as neither dual spaces nor wedge products are needed, and is invariably valid for spatial dimensions no greater than three. In addition, the fact that mimetic differences are not considered locally, but as defining linear operators, allows for an improved order of accuracy up to the physical domain boundary.

A numerical method for obtaining corresponding **discrete operators** which **mimics** the continuum differential **and** flux-integral operators which also allows for the same order of accuracy in the interior, as it does up to the spatial boundary of the physical domain, is also presented in this book. This method also generalizes those based on the Castillo–Grone discrete differential operators and utilizes the same spaces of grid functions for all three fundamental differential operators in a way that mimics the fundamental properties of the corresponding differential operators.

This book is organized as follows: In Chapter 1, we provide a brief overview of the various mimetic approaches, as well as many applications for which they have been used. We have also included some of the history behind the different methods, as well as a representative sample of the many contributors to the field. In Chapter 2, we review the use of continuum mathematical models as a way to motivate the natural use of mimetic methods. In Chapter 3, we present brief notes on numerical analysis as a way to facilitate the use of this book for courses on numerical methods that can be utilized for solving partial differential equations. In Chapter 4, we present mimetic differential operators in one-, two-, and three-spatial dimensions. Chapter 5 provides a comprehensive introduction to object-oriented programming and C++, with the intention of paving the way for readers uninitiated in this field. Chapter 6 provides an overview of the mimetic methods toolkit (MTK), our own API implementing Mimetic Discretization Methods. The MTK's purpose is to facilitate the development of computer codes in diverse computational frameworks, in order to solve arising problems through an intuitive and illustrative computational implementation of mimetic discretization methods. The API is available at:

http://www.csrc.sdsu.edu/mimetic-book/

In Chapter 7, we discuss the application of our mimetic methods to structured nonuniform meshes, and in Chapter 8, we present the following five different case studies:

1. Porous media flow

2. Carbon dioxide geologic sequestration

3. Electromagnetism

4. Seismic wave propagation

5. Geophysical fluid flow

Finally, as we move through the text, we will also present ways in which our Mimetic Methods Toolkit can be utilized to solve the equations we have reviewed in this book.

This text has been written primarily as a reference book; however, it can also be used as a textbook for any course involving the numerical solution of partial differential equations. It also works well in conjunction with other books on numerical partial differential equations, as the basic set of problems provided in the second text can be used as the starting point from which to find solutions for PDEs via the utilization of our method, as well as to compare with the alternative methods provided in the literature. A variety of sample problems are presented within this text, and many more will be posted on the book's web page:

http://www.csrc.sdsu.edu/mimetic-book/

Acknowledgments

This book represents our endeavor to compile the many concepts and results we have developed at the Computational Science Research Center (CSRC) over the years into an informative and comprehensive text. As it turns out, the most difficult aspect of writing this book has been finding a way to adequately acknowledge the many people whose contributions, both direct and indirect, have brought it to fruition. This has proven to be nearly impossible to accomplish, however, due to space constraints, so we must confine our thanks to a few people.

We begin by recognizing Mac Hyman, Mikhail Shashkov, and Stanly Steinberg; their invaluable work has been our primary influence, as our collaborations with each of them were essential to the conception of this book. Research undertaken by former and current graduate students of the CSRC is also a key component of this text, including the work of our Venezuelan group, which is led by Juan Guevara, Otilio Rojas, and German Larrazabal. This team has been a long-standing contributor to our endeavors via their work on the application of our mimetic ideas, as well as their assistance in organizing important mimetic meetings in South America.

Our thanks also go out to Mark Yasuda, Carlos Bazan, Peter Blomgren, and Ali Nadim for reading and providing valuable feedback on various sections of this book. We would also like to thank Mohammad Abouali for his contribution to the case studies included in the text. Our gratitude also goes out to Jessica Nombrano Larsen for her help with editing the book. We would also like to thank Diana Sanchez for her contribution to some of the figures and to the cover design. Lastly, we would like to thank Eduardo Sanchez for his contributions to the case studies, as well as for his help with the book's figures, LaTeX editing, and his role as lead designer and programmer of the Mimetic Methods Toolkit (MTK), which will be introduced in detail within this book.

Chapter 1

Introduction

This book provides the required elements for applying modern techniques to the discretization of a wide range of continuum mechanics problems. Mimetic or compatible methods begin by discretizing the vector calculus operators used to describe continuous problems. They then use the resulting discrete operators to discretize the given problem. If the continuous problem is stable, the mimetic discretization will typically result in a stable numerical scheme. However, although stability is a desirable characteristic of any numerical method, it is usually very difficult to accomplish, especially in regard to high-order discretization. Many classical discretization approaches begin by discretizing a given problem to some specified order of accuracy, followed by an attempt to prove the method's stability, which can be a very difficult task. In contrast, mimetic methods automatically preserve many important properties of the continuous problem (e.g., conservation laws), thereby contributing to the stability of the method.

Continuum models are used in many sciences, and their relevant magnitudes are usually described as unknown **distributions** or fields, which are dependent upon time and space in the **continuous domain** under consideration. If such fields are piecewise continuous or locally integrable dependent variables, then the conservation laws of the continuum models can be expressed using the integral forms of the underlying **conservation principle**.

Mimetic methods are used in general logically rectangular grids [2], [3]; irregular or unstructured grids [4], [5], [6], [7]; triangular grids [8], [9], [10]; and polygonal grids [11], [12], [13]. They can be made higher order in 3-D [14], as well as monotone [15]. Mimetic spectral and pseudo-spectral methods have also been derived [16], [17], [18], and have been implemented in a multiscale setting [19] using interpolation [20]. Iterative solvers have been used to invert the matrices arising from mimetic discretization [21]. There are many applications of mimetic methods in solving continuum problems, including in the geosciences (porous media) [19], [22], [23], [24], [25], [26], [27], [28]; fluid dynamics (Navier-Stokes, shallow water) [29], [30], [31]; image processing [32], [33], [34], [35]; general relativity [2]; and electromagnetism [36].

Mathematical models of continuum mechanics problems are typically described by boundary value problems, and expressed as either a system of partial differential equations (PDEs) or as integral equations. To facilitate their numerical solution, these equations can be discretized by any one of a large number of techniques; standard methods include various finite difference

1

and finite element approaches. These traditional methods are often applied by discretizing the defining system of equations directly. One disadvantage of such an approach is that the discretization scheme selected may have little connection with the underlying physical problem. Mimetic methods, on the other hand, begin by first discretizing a problem's underlying continuum theory. By "discretizing the continuum theory," we mean that mimetic methods initially construct a discrete mathematical analog of a relevant description of continuum mechanics (with the description usually taking the form of a physical conservation or constitutive law). The discrete form of these conservation or constitutive law constrains the structure that discrete operators can take. After building discrete operators that obey the discretized physical law, these mimetic operators can then be substituted into a system of partial differential equations or integral equations. This yields a mimetic discretization for the boundary value problem, which automatically satisfies the discrete version of the physical law on the corresponding domain under consideration. As a result, discretizations obtained using mimetic methods tend to replicate much of the behavior found in the actual continuum problem. Since the physical laws are, in effect, built into the discretization, mimetic methods turn out to be good candidates for modeling even the most challenging problems, such as those involving anisotropic or strongly inhomogeneous material properties.

The physical basis for mimetic discretizations also helps to reduce the occurrence of various nonphysical numerical artifacts that can occur when using a traditional discretization technique. Examples of such artifacts include

1. Oscillatory solution contaminants (also known as hourglass modes);

2. Nonphysical spectra for Laplacian operators, when dealing with elliptic partial differential equations; and

3. Late-time instabilities in which errors increase rapidly and significantly over time.

To the extent that physical conservation laws forbid such numerical side-effects from occurring in the real world, they are often avoided by mimetic discretizations that have incorporated the discrete analogs of these governing constraints. Given a discrete conservation law and a staggered grid (uniform or nonuniform), one can readily create mimetic approximations of high order in the interior of a region. Building mimetic operators with comparably high-order approximations at or near the boundary of a region, however, is considered challenging, even in the simplest case: namely, that of a uniform, one-dimensional grid.

To date, the search of mimetic or compatible discretizations for use in the solution of PDEs has been met with varying degrees of success. There are many researchers currently active in this area; all pursuing different approaches to achieve the same goal. A case in point is the famous Yee grid [37] (published in 1966) for discretizing Maxwell's equations, which is a prime example of a mimetic finite difference method. More recent developments include the finite difference time domain (FDTD) [38], [26] and finite integration theory method (FIT) [39], [40], [41] as well as the work of M.E. Rose [42] and Blair Perot and his collaborators [43], [44], [45], [46], [47], who have also developed discretizations based on mimetic ideas. Finite volume methods such as those used in [48] to study conservation laws use mimetic ideas to discretize the divergence operator or integral forms of the conservation laws. Support-operator methods have been used to model diffusion [36], [49], [49], [50], [51], electromagnetism [52], [36], and hydrodynamics [53].

While we have endeavored to be comprehensive in our bibliography, there is a great body of work in the scientific literature regarding the concept of mimetic discretizations, so it is likely incomplete. The work of J.M. Hyman is particularly worth mentioning, as he established an important group at Los Alamos National Laboratory, and was the first to use the word *mimetic* in association with what others call "compatible discretizations" [22], [54], [36], [55], [56], [57], [58], [59], [60], [61], [62], [63] [52], [64]. Dr. Hyman's group at Los Alamos also included Mikhail Shashkov [50], [51], [65], who was previously a member of A.A. Samarskii's group, where the so-called "support operators method" had its genesis [66], [67], [68], [69]. Professor Stanly Steinberg of the University of New Mexico maintains a webpage listing approximately 50 researchers currently working on mimetic ideas, and has also directed students doing their dissertations in this area [70], [71]. And, largely due to the pioneering efforts of Professor Juan Guevara [72], [73] from the Universidad Central de Venezuela, there has also been sustained activity in mimetic research in Venezuela, including the hosting of three important meetings devoted to mimetic methods and applications.

Other significant contributions have come from the electromagnetic community, with Nédélec [74], [75]; Teixeira [76], [77], [78], [79]; Bossavit [80], [81], [82], [83], [84], [85], [86], [87], [88], [89], [90], [39]; and Mattiussi [15], [91], [92], [93] being important researchers in this field. The Uppsala group has also had a significant impact on the field, having pioneered the summation by parts method [94], [95], [96], [97], [98], [99], which originally began with the work of Godela Scherer's PhD thesis, under the supervision of H.O. Kreiss [100]. However, the Uppsala group uses a collocated or nodal grid in their work, which produces only one operator, and the order of accuracy at the boundary is lower than in the interior of the principal domain. The Castillo-Grone method differs from the Uppsala method in its utilization of a staggered mesh. In addition, it achieves the same high-order accuracy in the interior as it does at the boundary with a diagonal weighted inner product.

There are also researchers coming from the finite element, mixed finite element, finite volume, topology, and algebraic topology camps currently using mimetic ideas. Papers have also been written dealing with convergence and error estimates as well as stability. In fact, researchers coming from many different areas are now converging on the types of key properties desired for numerical methods associated with mimetic discretizations.

A mimetic, one-variable calculus using dual grids is presented in [101], to discretize general Sturm-Liouville boundary-value problems. Estimates of the truncation error in mimetic methods are derived [102], and estimates of the error in the solution and its derivative are also given. The paper [64] extends these ideas to rough logically rectangular grids, and proves that solutions of the mimetic discretization of the scalar Laplacian converge to a solution of the continuum Laplacian. The paper [103] provides proof of convergence that quantifies the dependence on the roughness of the grid systems. An alternative proof of convergence based on finite element methods is given in [104], [105], while super-convergence results are given in [106].

Ideas from vector calculus, differential geometry, and algebraic topology have been used to create, understand, and analyze discretization methods [107], [108], [109], [23], [110], [15], [92], [93], [45], [111], [77], [112], and have also been used to build high-level programming languages for discretization [113], [114], [115], [116], [117], [118].

In [119], the completion to involution of discretizations is studied. Compatible finite element methods [120], [121] are somewhat similar to (generalized) finite difference mimetic schemes. Compatible schemes are commonly derived using ideas from differential geometry and algebraic topology [121], [122], [90], [39], [123], [124], [125], [125], [126], [127].

Researchers in electromagnetism have made extensive use of compatible schemes [128], [129], [80], [81], [82], [84], [85], [86], [87], [88], [89], [130], [131], [132], [133], [134], [62], [135], [75], [136], [137], [76], [78], [138], [139], [140], with much emphasis being placed on Hodge star operators and Whitney forms [127]. Such convergence of finite difference and finite element methods can also be seen in the field of computational aerodynamics.

Although mimetic discrete differential operators are linear, they can also be used effectively to solve nonlinear partial differential equations. The main difference is that the resulting algebraic system of equations is no longer linear. In cases such as these, there are many algorithms that can be used to solve it. For instance, see [35] for their use in image processing, where a mimetic algorithm was implemented to solve a diffusion equation arising from image restoration models. In fact, this algorithm outperformed standard discretization methods previously used within the image processing community.

Since it is seldom possible to find exact, explicit solutions to the more challenging real problems, **numerical methods** are often used. These methods are based upon **discrete models** which reproduce or simulate important properties of the continuum models under consideration, with a special emphasis on the relevant conservation laws.

Simply stated, the **numerical solution** of a functional equation consists of a finite set of numbers from which the distribution of the unknown piecewise continuous field function u in the domain of interest can be approximately obtained.

Discretizing the distribution of u is the starting point of any **discretization numerical method**, resulting in a **grid function** $\tilde{u}$, whose values are considered at a finite number of locations or **grid points** $(x_i, i = 1, ..., M)$. When the values of $\tilde{u}$ are sequentially arranged, they constitute an M-vector whose i-th component is $u_i^* = u(x_i)$. The grid points belong to the Euclidean vector space E^d, whose dimension d equals the number of independent variables forming the arguments of u, so that i stands for a multi-index. When $d \geq 2$, the set G of grid points or **grid**, may look like a lattice in E^d. Closely associated with the grid G, one can define a **mesh** or set of d-dimensional **cells**. For example, in the case of $d = 3$, the cell or mesh elements could be convex or nonconvex polyhedra and the grid points could be cell-centered, face-centered, edge-centered, or vertices. A single mesh can be associated with more than one grid G, as in the case of **staggered grids**. Although we only considered structured grids in this book, keeping in mind applications in parallel computing, the method presented can also be extended for use with unstructured grids.

The next step in the numerical method consists of **adequately** defining another grid function, $u^h = (u_1^h, ..., u_M^h)$, where h stands for a typical **mesh size**. The smaller h is, the finer the mesh looks, as M increases with diminishing h. The method must provide some **scheme** resulting in a system of M algebraic equations for the unknown u_i^h, $i = 1, ..., M$, and must also exhibit an **algorithm** solving this system. The **adequacy** of the definition for u_i^h rests upon the **convergence** of the **discrete scheme**, in which u_i^h is so defined that it becomes a good approximation for $\tilde{u}_i$, or, more precisely, when $\lim(u^h - \tilde{u}) = 0$ as h tends to 0. Notice that for each h, both u^h and $\tilde{u}$ belong to the same M-dimensional vector space, where the size of their difference can be evaluated using an appropriate norm (whether Euclidean or otherwise).

Many of these situations are best treated by also discretizing the physical domain, thereby allowing the use of different **profile assumptions** in each subdomain (e.g., exponentials rather than the usual Taylor-induced polynomials). Therefore, a general discretization method is completed with the use of **piecewise profiles** to evaluate the required **approximate value of integrals**, or the **difference approximations for derivatives**.

A large number of partial differential equations (PDEs) are formulated using divergence, gradient, and curl **first-order differential operators**, which all admit a common, **coordinate-free formulation** by means of the DEL operator, whose formulation is given as a **limiting flux per unit differential volume**.

Maxwell's equations can be combined in such a way that both the electric field E and the magnetic induction field B (assumed to be twice differentiable) satisfy a vector wave equation where the spatial variation is described by the operator *grad div − curl curl*. This is a typical example of wave-yielding PDEs of **hyperbolic type**, and exhibits the composition of the fundamental first-order differential operators.

We will also consider the steady state fields $E = grad\ V$, $B = grad\ U$, where the scalar potentials U and V satisfy *div grad U = 0* (Laplace's equation), and *div grad V = −f* (Poisson's equation), where f stands for the continuous electric charge density. These two equations are the canonical PDEs of the **elliptic type** (they are homogeneous and nonhomogeneous, respectively), and are subject to Dirichlet, Neumann, and Robin-type boundary conditions (BCs).

Additionally, we will examine problems arising in heat conduction and diffusion processes where the time rate of change of temperature T is proportional to $div\,(\mathcal{K}\,grad\,T)$, and the tensor $\mathcal{K}$ measures the medium heat conductivity (this is the representative form of PDEs of **parabolic type**). Changing the unknown function T to concentration C, and the tensor conductivity parameter $\mathcal{K}$ to a tensor diffusivity parameter $\mathcal{D}$, results in a PDE governing diffusion of some substance.

In this book we present a mimetic method wholly based upon Gauss' divergence theorem rather than upon the general Stokes' theorem, which is simpler (with no dual spaces or wedge products needed), and which is always valid for spatial dimensions no greater than three. Furthermore, the fact that mimetic differences are not considered locally but rather as defining linear operators allows for the improvement of the order of accuracy up to the physical domain boundary.

Chapter 2

Continuum Mathematical Models

A road map is used to guide a driver from one location to another. Such a map is a much-simplified representation of the actual physical reality connecting both locations, yet; it helps the driver achieve a goal.

Mathematically ideal objects can be used analogously, since:

1. Points can model tiny particles.

2. 1-D lines, whether straight or curved, can be used to model thin wires, guitar strings, rods, capillary blood vessels, and even transcontinental ballistic missile trajectories;

3. 2-D surfaces provide good representations of thin plates, membranes, or sheets; and

4. 3-D solids can be used to describe the **physical domain** in 3-D space, where some material substance, body, or physical field is considered to be present at a certain point in time.

Instead of treating mass particles or electric charges as points, for convenience, it is common to consider them to be distributed continuously in space. The physical fields involved are also assumed to be continuously distributed throughout the domain under consideration, but their corresponding continuous mathematical fields can be of different types, due to the complexity of the physical magnitudes in play.

In most cases, these mathematical fields are piecewise continuously differentiable and can be

1. **Scalar-valued**, such as pressure, mass-density, electric charge density, temperature, electrostatic or gravitational potential, etc.

2. **Vector-valued**, including gravitational, electric or magnetic fields; velocity fields describing fluid motion; potential gradients; and others. Such vector fields are denoted by boldfaced Latin letters, such as $\mathbf{v}$, $\mathbf{E}$, and $\mathbf{B}$. In the case of a given grid function $\mathbf{u}$, its components would be denoted by single-indexed letters such as u_i, unless time steps should also be considered; in which case, a superscript would also be added; e.g., for the n-th time step, we would write: u_i^n.

3. **Tensor-valued**, such as the stress tensor defined for elastic bodies, or the energy-momentum tensor of an electromagnetic field. Tensor fields are denoted by calligraphic letters, such as $\mathscr{T}$, Σ, and their components by double-indexed letters such as T_{ij} and σ_{ij}. Whenever non-Cartesian coordinates are used, a distinction between covariant, contravariant, and mixed tensors will also appear.

2.1 Physically Motivated Mathematical Concepts and Theorems

2.1.1 Flux and Flux Density

Let M be the mass of a substance **at rest** within a region Ω of space. Assume that the substance is homogeneous and continuously distributed in Ω, a 3-D solid with volume V (in 1-D or 2-D, Ω becomes a surface σ or a curve C, respectively).

The familiar concept of **mass density** ρ relates geometrical magnitudes to physical ones by means of the product rule:

$$M = \rho V. \tag{2.1}$$

When the substance exhibits inhomogeneities, a density or mass distribution scalar function, $\rho : \Omega \longmapsto \mathbb{R}$, yields M by means of the volume integral:

$$M(\Omega) = \int_{\Omega} \rho \, dV. \tag{2.2}$$

Notice its similarities with the homogeneous case, since dV can be thought of as a multiplying "factor" when one recalls the Riemann sum approach to computing definite integrals.

Suppose now that a velocity field $\mathbf{v}$ is instantaneously imparted to all particles within Ω.

Assume that the boundary of Ω is a closed piecewise smooth surface σ, so that σ separates the space into an inner region Ω and an outer region that extends to infinity.

Loosely speaking, *smoothness* means that it is possible to define on σ a piecewise continuous field of normal unit vectors $\mathbf{n}$ (which have been outwardly directed for convenience).

Let σ^+ be the subset of σ, where the particles will exit from; Ω, i.e., $\mathbf{v} \cdot \mathbf{n} > 0$, and let σ^- be the subset of σ, where the particles appear to be just entering Ω, so that $\mathbf{v} \cdot \mathbf{n} < 0$ throughout σ^- (Figure 2.1). Obviously, $\sigma = \sigma^+ \cup \sigma^- \cup \Gamma$, with Γ being the subset of σ and the imparted velocity $\mathbf{v}$ is tangent to σ, or, equivalently, where $\mathbf{v} \cdot \mathbf{n} = 0$.

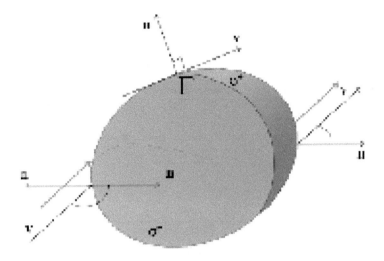

Figure 2.1: Velocity flow across σ.

Now, let us define the **net mass outflux** through surface σ as the difference between the outflux and the influx; that is,

$$\int_{\sigma^+} \rho \mathbf{v} \cdot \mathbf{n} \, d\sigma - \int_{\sigma^-} \rho \mathbf{v} \cdot (-\mathbf{n}) \, d\sigma = \int_{\sigma} \rho \mathbf{v} \cdot \mathbf{n} \, d\sigma = \int_{\sigma} \rho \mathbf{v} \cdot (\mathbf{n} \, d\sigma). \qquad (2.3)$$

Here, the mass flux through σ means mass flow crossing a surface σ per unit differential time, with $\rho \mathbf{v}$ standing for the **mass vector flux density**, and $\mathbf{n} d\sigma$ is the multiplying "factor."

2.1.2 Gauss' Divergence Theorem

This fundamental formula, with a heuristic deduction given in Appendix A, reads as follows:

$$\int_{\sigma} \rho \mathbf{v} \cdot \mathbf{n} \, d\sigma = \int_{\Omega} \nabla \cdot (\rho \mathbf{v}) \, dV. \qquad (2.4)$$

Equation (2.4) relates the flux (defined as some surface integral) to a volume integral. It also introduces us to our most important first-order spatial differential operator, called the **divergence operator** (denoted by div or $\nabla \cdot$). Gauss' formula is useful for obtaining the mathematical expressions of physical **conservation laws**, such as the continuity equation, which expresses mass conservation (see §2.3.1).

In Cartesian coordinates x, y, z, and for $\mathbf{v} = P\mathbf{i} + Q\mathbf{j} + R\mathbf{k}$, we define[1]:

$$div\ \mathbf{v} \triangleq \nabla \cdot \mathbf{v} = \frac{\partial P}{\partial x} + \frac{\partial Q}{\partial y} + \frac{\partial R}{\partial z}. \tag{2.5}$$

As a result, *div* changes vector-valued functions into scalar-valued functions, and can be generalized in such a way that it turns second-order tensors into vectors (see Appendix B).

When this operator is applied to the product $\rho\mathbf{v}$, we obtain

$$div(\rho\mathbf{v}) = \rho\ div\ \mathbf{v} + \mathbf{v} \cdot grad\ \rho. \tag{2.6}$$

In Cartesian coordinates:

$$grad\ \rho \triangleq \nabla\rho = \frac{\partial \rho}{\partial x}\mathbf{i} + \frac{\partial \rho}{\partial y}\mathbf{j} + \frac{\partial \rho}{\partial z}\mathbf{k}. \tag{2.7}$$

Now, *grad* is called the **gradient operator**, and is our second first-order differential operator (commonly denoted as ∇).

When no confusion arises, we write

$$\frac{\partial \rho}{\partial x} = \rho_x, \quad \frac{\partial \rho}{\partial y} = \rho_y, \quad \frac{\partial \rho}{\partial z} = \rho_z. \tag{2.8}$$

Note that, contrary to *div*, *grad* can change scalar functions into vector functions. So, in its **extended form**, Gauss' divergence theorem reads, as follows:

$$\int_\Omega \rho\ div\ \mathbf{v}\ dV + \int_\Omega \mathbf{v} \cdot grad\ \rho\ dV = \int_\sigma \rho\mathbf{v} \cdot (\mathbf{n}\ d\sigma). \tag{2.9}$$

If Ω is reduced to $[0,1]$, then the theorem returns to the **integration by parts (IBP)** version of the fundamental theorem of calculus (FTC), such that

$$\int_0^1 \rho\frac{dv}{dx}\ dx + \int_0^1 v\frac{d\rho}{dx}\ dx = \rho(1)v(1) - \rho(0)v(0). \tag{2.10}$$

If we consider the real line segment $[0,1]$ as a 1-D model for a blood capillary vessel, with $\mathbf{v} = v\mathbf{i}$, then, after taking a cross section of the vessel so as to have a very small unit area, we get

$$M([0,1]) = \int_0^1 \rho\ dx, \quad \mathbf{n}(1) = \mathbf{i}, \quad \mathbf{n}(0) = -\mathbf{i}, \tag{2.11}$$

and the net mass outflux is

$$(\rho\mathbf{v})(1) \cdot \hat{i} + (\rho\mathbf{v})(0) \cdot (-\mathbf{i}) = \rho(1)v(1) - \rho(0)v(0). \tag{2.12}$$

---

[1]The symbol $\triangleq$ stands for "equal by definition."

When the scalar function models a given pressure p instead of a mass density ρ, $grad\ p$ has a different physical meaning, thus allowing the computation of the global surface pressure force:

$$\mathbf{F}(\sigma) = -\int_{\sigma} p(\mathbf{n}\ d\sigma) = -\int_{\Omega} grad\ p\ dV. \tag{2.13}$$

This equation is readily obtained using Gauss' formula three times (see Appendix C).

2.1.3 Work, Circulation, and Electromotive Force

When there is a vector field $\mathbf{F}$, representing physical forces acting along some direct path from point A to point B, then the work W_{AB} done by $\mathbf{F}$ upon a particle (forcing it to move from A to B), is defined as a line integral:

$$W_{AB} \triangleq \int_{A}^{B} \mathbf{F} \cdot d\mathbf{r}, \tag{2.14}$$

with $d\mathbf{r} = dx\mathbf{i} + dy\mathbf{j} + dz\mathbf{k}$ in Cartesian coordinates. If the directed path AB becomes a positively oriented **closed** curve C, and we interpret the field $\mathbf{F} = \mathbf{v}$ as a given fluid velocity, then the **circulation** of $\mathbf{v}$ is

$$\int_{C} \mathbf{v} \cdot d\mathbf{r}, \tag{2.15}$$

and, if $\mathbf{F} = \mathbf{E}$, (the electric field), then the electromotive force (EMF) can be defined, as follows:

$$\int_{C} \mathbf{E} \cdot d\mathbf{r}. \tag{2.16}$$

Analogically, if $\mathbf{F} = \mathbf{B}$ (the magnetic induction field), then the magnetomotive force (MMF) can be defined, as follows:

$$\int_{C} \mathbf{B} \cdot d\mathbf{r}. \tag{2.17}$$

If $\mathbf{F} = grad\ U$ for a scalar-valued function U, and

$$grad\ U \cdot \frac{d\mathbf{r}}{dt} = \frac{dU}{dt}, \tag{2.18}$$

then

$$\int_{A}^{B} grad\ U \cdot d\mathbf{r} = U(B) - U(A). \tag{2.19}$$

2.1.4 Stokes' Theorem and *curl*

If C is a closed, smooth curve bounding any curved surface σ, then Stokes' theorem states that the electromotive force can be defined as

$$\int_C \mathbf{E} \cdot d\mathbf{r} = \int_\sigma \mathbf{n} \cdot curl\ \mathbf{E}\, d\sigma. \tag{2.20}$$

Note that C (which should be thought of as lying on a floor) is oriented in such a manner that, when traversed by an observer, the points of σ remain to his left. Having fixed this orientation for the bounding curve C, the orientation of σ now follows the rule of the screwdriver, such that the unit normal $\mathbf{n}$ to σ points in the same direction as the advancing screwdriver. (An argument for a quick mathematical understanding of Stokes' formula is given in Appendix D.)

We now have our third first-order spatial differential operator: the *curl*, which is also called the **rotor operator**. In Cartesian coordinates, it is given as

$$curl\ \mathbf{v} \triangleq \nabla \times \mathbf{v} = \begin{vmatrix} \mathbf{i} & \mathbf{j} & \mathbf{k} \\ \frac{\partial P}{\partial x} & \frac{\partial Q}{\partial y} & \frac{\partial R}{\partial z} \\ P & Q & R \end{vmatrix}. \tag{2.21}$$

Note that the *curl* operator sends vector functions into vector functions.

The *curl* operator relates to the *grad* operator in several ways. For example: when $\nabla \times \mathbf{E} = 0$, throughout the continuous domain of $\mathbf{E}$ (these vector fields are called **conservative**), because, in that case

$$\int_C \mathbf{E} \cdot d\mathbf{r} = 0, \tag{2.22}$$

for any closed curve C, implying that

$$\int_A^B \mathbf{E} \cdot d\mathbf{r}. \tag{2.23}$$

does not depend upon the path followed in order to go from A to B; therefore, a scalar-valued function U, called **the potential for E**, must exist, and it is defined at a variable point B for a fixed point A, by

$$U(B) = \int_A^B \mathbf{E} \cdot d\mathbf{r}. \tag{2.24}$$

For such a conservative $\mathbf{E}$:

$$grad\ U(B) = \mathbf{E}(B). \tag{2.25}$$

It can also be stated as

$$curl\ \mathbf{E} \equiv 0, \tag{2.26}$$

or, $\mathbf{E}$ "conservative," which implies there exists a potential U, such that (2.25) holds.

Other relations between *grad* and *curl* include the vector identities

$$curl\ grad\ U \equiv 0, \tag{2.27}$$

and

$$curl\ curl\ \mathbf{E} \equiv 0. \tag{2.28}$$

Physically, the *curl* of the fluid velocity field, called the **vorticity** of the corresponding fluid flow, is germane to the fluid's viscosity properties.

A strong grasp of the physical meaning of *curl* $\mathbf{v}$ can be attained with the velocity field of an infinite vortex line along the z-axis, within a viscous incompressible fluid, such that

$$\mathbf{v} = \omega\mathbf{k} \times (x\mathbf{i} + y\mathbf{j}) = \omega(x\mathbf{j} - y\mathbf{i}), \tag{2.29}$$

which clearly satisfies

$$div\ \mathbf{v} = 0, \tag{2.30}$$

$$curl\ \mathbf{v} = 2\omega\mathbf{k}, \tag{2.31}$$

and

$$\mathbf{k} \cdot curl\ \mathbf{v} = 2\omega, \tag{2.32}$$

so that, in this very simplified model of a vortex within an infinite fluid, we see that *curl* $\mathbf{v}$ has the direction of the axis of rotation, with a magnitude of twice the angular velocity ω.

A more realistic model features a nonzero *curl* $\mathbf{v}$ (exemplified by a finite volume fluid), and is provided by an incompressible inviscid liquid mass which partially fills a vertical, cylindrical container, whose axis is coincident with the z-coordinate axis. This liquid, initially at rest, fills the container to a height H; when the cylindrical container is forced to rotate around its axis with constant angular velocity ω, the liquid within also rotates, much as a rigid body would. The liquid free surface deviates from the original static shape (a horizontal plane), and looks like a paraboloid of revolution after it has attained a steady state. We can now assume that the points at the liquid free surface experience atmospheric pressure throughout, since the height variations of the surface are small when compared with atmospheric height. This constant reference pressure can be taken to be zero, and, as before, $\mathbf{v} = \omega\mathbf{k} \times (x\mathbf{i} + y\mathbf{j}) = \omega(x\mathbf{j} - y\mathbf{i})$, so that *curl* $\mathbf{v} = 2\omega\mathbf{k}$.

As a result of the incompressibility assumption, the liquid density is some constant ρ. Now, when we use the cylindrical coordinates (r, θ, z), there is a dynamical pressure $p(r, z)$ created, due to the centrifugal force generated by the circular motion of the small volume element within the fluid. This force then satisfies

$$-\rho\frac{d\mathbf{v}}{dt} = \rho\omega^2(x\mathbf{i} + y\mathbf{j}). \tag{2.33}$$

Here, the conservation of momentum, when applied to a rotating differential element of liquid, yields

$$\rho\frac{d\mathbf{v}}{dt} = -\rho g\mathbf{k} - \nabla p(r, z). \tag{2.34}$$

Now, let h denote the maximum height reached by the fluid above the initial static surface plane, and l the depth reached by the lowest point on the free surface of the rotating liquid; as you can see in Appendix E:

$$h + l = \frac{\omega^2}{2g}a^2, \tag{2.35}$$

where a is the radius of the cylinder. As the liquid is incompressible, we can separate the values h and l, since the volume under the paraboloid's free surface must equal the quantity $\pi a^2 H$.

With numerical methods, the real-world shape of free surfaces can be approximated with high accuracy, one example being ocean waves hitting ships and then bouncing back again.

However, it is incorrect to conclude that the streamlines must be curved in order to have a nonzero *curl*. **Poiseuille's flow** provides a simple example of a velocity field $\mathbf{v}$ having a nonzero *curl* $\mathbf{v}$, despite the fact that its streamlines are a family of parallel **straight** lines, as it is laminar (nonturbulent) steady viscous flow, which, at low velocities in a cylindrical tube, is subject to a given constant pressure gradient P along its axis. By also using the cylindrical coordinates (r, θ, z), with the positive z-axis coincident with the cylinder's axis as well as with the direction of flow, it can be shown (see Appendix F) that

$$\mathbf{v}(r) = -\frac{1}{4\eta}P(R^2 - r^2)\mathbf{k}, \tag{2.36}$$

where η is the dynamic fluid viscosity, $r^2 = x^2 + y^2$, R is the cylinder's radius, and

$$P = \frac{dp}{dz}. \tag{2.37}$$

The field $\mathbf{v}$ does not depend upon either z or θ, so that $\mathbf{v}(r) = v(r)\mathbf{k}$. We can now assume that $v(R) = 0$ ("no slip" boundary condition), and that

$$\frac{dv}{dr}(0) = 0 \tag{2.38}$$

(the maximum velocity at the center of the tube).

Since $r^2 = x^2 + y^2$, and, assuming P and η are such that

$$-\frac{P}{4\eta} = 1, \quad P < 0, \tag{2.39}$$

we find that

$$(curl\ \mathbf{v})(x,y) = -2y\mathbf{i} + 2x\mathbf{j} = 2\mathbf{k} \times (x\mathbf{i} + y\mathbf{j}). \qquad (2.40)$$

2.1.5 2-D Green's Formula

This well-known result reads, as follows:

$$\iint\limits_{\Omega} \left(\frac{\partial Q}{\partial x} - \frac{\partial P}{\partial y}\right)(x,y)\ dxdy = \int\limits_{C} P(x,y)dx + Q(x,y)\ dy. \qquad (2.41)$$

This formula, as well as the 3-D formulae known as Green's identities, are proved in Appendix G.

2.1.6 Laplace's Equation and 2-D Incompressible and Inviscid Flows ("Potential Flows")

As can be seen in Appendix H, the time rate of change of the fluid volume $V(t)$ inside a 3-D region Ω at time t is related to $div\ \mathbf{v}$ by

$$\frac{d}{dt}V(t) - \int\limits_{\Omega} \nabla \cdot \mathbf{v}\ dV = 0. \qquad (2.42)$$

Therefore, if the fluid is incompressible, then $\frac{d}{dt}V(t) \equiv 0$, and Ω being arbitrary, it is implied that, for all time t, $\nabla \cdot \mathbf{v} \equiv 0$ in Ω for continuous $div\ \mathbf{v}$.

However, if the fluid has no viscosity, then $curl\ v = 0$ in Ω, so that there exists some scalar-valued function U such that $\mathbf{v} = grad\ U$; therefore, $\nabla \cdot \nabla U \equiv 0$ in Ω, and U satisfies Laplace's equation $\Delta\ U = 0$ in Ω, where $\Delta = \nabla \cdot \nabla$ is **Laplace's operator.**

If Ω is a 2-D region, then $\nabla\ U(x,y) = U_x(x,y)\mathbf{i} + U_y(x,y)\mathbf{y}$, so that we must have Laplace's equation:

$$\nabla \cdot \nabla\ U(x,y) = (\frac{\partial}{\partial x}U_x + \frac{\partial}{\partial y}U_y)(x,y) = (U_{xx} + U_{yy})(x,y) \equiv 0 \text{ in } \Omega. \quad (2.43)$$

This allows us to establish a connection between scalar-valued functions U, thereby satisfying a 2-D Laplace's equation, together with the associated vector velocity field $\mathbf{v} = grad\ U$, and the analytic complex functions of a complex variable $z = x + iy$. As before, we denote $x - iy = \bar{z}$.

Let $f(z) = U(x,y) + iW(x,y)$ be a differentiable complex function, so that

$$f'(z) = (U_x + iW_x)(x,y) = (W_y - iU_y)(x,y). \qquad (2.44)$$

This implies that $U_x = W_y$, $W_x = -U_y$, so that $\overline{f'(z)} = (U_x + iU_y)(x,y)$ is the complex function naturally associated with the 2-D vector valued function $\mathbf{v} = grad\ U$ evaluated at (x,y) and $|\overline{f'(z)}| = ||\nabla U(x,y)||$. Analogously, this formula shows that f acts as a form of complex potential for the complex velocity field $(U_x + iU_y)$. (In fact, it is the exact analog of a potential, except for the complex conjugation.)

Also, note that $U_{xx} = W_{xy} = -U_{yy}$, and $\nabla \cdot \nabla \, U(x,y) \equiv 0$, so that the complex conjugates of complex first derivatives can model incompressible inviscid 2-D velocity flows.

Now consider a directed path AB in the complex z-plane. The complex integral is written as

$$\int_A^B f'(z) \, dz = f(B) - f(A) = U(B) - U(A) + i[W(B) - W(A)], \qquad (2.45)$$

when the path AB does not cross a discontinuity point for f.

Since $dz = dx + i dy$, then $f'(z)dz = U_x dx + U_y dy + i[U_x dy - U_y dx]$ and

$$\int_A^B f'(z) \, dz = \int_A^B \nabla U(x,y) \cdot d\mathbf{r} + i \int_A^B \nabla \, W(x,y) \cdot d\mathbf{r} \qquad (2.46)$$

$$= \int_A^B \nabla U(x,y) \cdot d\mathbf{r} + i \int_A^B \nabla \, U(x,y) \cdot \mathbf{n} \, ds, \qquad (2.47)$$

where $ds = |d\mathbf{r}|$ and $\mathbf{n}$ is the unit normal vector from left to right of the directed path AB, so that the real and imaginary parts of

$$\int_A^B f'(z) \, dz \qquad (2.48)$$

are real-valued line integrals from A to B with **a physical meaning** of

$$\int_A^B \nabla U(x,y) \cdot d\mathbf{r} = \int_A^B \mathbf{v} \cdot d\mathbf{r} = U(B) - U(A). \qquad (2.49)$$

Therefore, the real part of

$$\int_A^B f'(z) \, dz \qquad (2.50)$$

measures the difference of the potentials along the path

$$\int_A^B \nabla U(x,y) \cdot \mathbf{n} \, ds = \int_A^B \mathbf{v} \cdot \mathbf{n} \, ds. \qquad (2.51)$$

In a related fashion,

$$\int_A^B \nabla U(x,y) \cdot \mathbf{n} \, ds = W(B) - W(A) \tag{2.52}$$

implies that the imaginary part W of $f(z)$, called the **streamline function**, is constant along directed paths which coincide with "streamlines," i.e., lines everywhere **tangent** to the velocity field $\mathbf{v}$. Lines of constant U are called **equipotentials**.

As a result, the imaginary part measures the **flux** of $\mathbf{v}$ across the path AB from left to right.

When the vector field $\mathbf{v}$ represents force fields, then the streamlines are called "lines of force." However, we generally speak more of **flux lines**, which, in the case of potential flows, are orthogonal to the equipotential lines since $U_x W_x + U_y W_y \equiv 0$.

If the directed path is a closed curve C, then $U(B) - U(A) = 0$; but, if $A = B$ is a discontinuity point for W, then the flux

$$\int_C \nabla U(x,y) \cdot \mathbf{n} \, ds \tag{2.53}$$

will differ from 0. This happens when: $z = x + iy = rcis(\theta)$, $-\pi < \theta \leq \pi$, $f(z) = \ln(r) + i\theta = U + iW$, since for $W(r,\theta)$, it holds that $W(1,\pi) = \pi$, and

$$W(1,-\pi) = \lim_{\theta \to -\pi} W(1,\theta) = -\pi, \tag{2.54}$$

so that $W(B) - W(A^-) = 2\pi$.

This provides a physical understanding, **in terms of the flux notion** for the known formula

$$\int_C \frac{1}{z} \, dz = 2\pi i, \tag{2.55}$$

where C is a circumference centered at the origin, with radius r and a counter-clockwise orientation, as $\frac{1}{z} = f'(z)$ for $f(z) = \ln(z) = \ln(r) + i\theta$, which is discontinuous on the negative real axis. When we use $dz = ircis(\theta) \, d\theta$, $ds = r \, d\theta$, then

$$\int_C \frac{1}{z} \, dz = \int_{-\pi}^{\pi} i \, d\theta = 2\pi i = i \int_C \frac{1}{r} \, ds = i \int_C \left| \frac{1}{\bar{z}} \right| \, ds = i \int_C \|\nabla U\| \, ds. \tag{2.56}$$

We observe that

$$\overline{f'(z)} = \frac{1}{\bar{z}} = \frac{z}{r^2} = \frac{cis(\theta)}{r} \tag{2.57}$$

has the same direction as $cis(\theta)$ or $\mathbf{n}$, so that $\mathbf{n} \cdot \nabla U(x,y) = \|\nabla U\|$.

These complex calculus considerations are useful as a means of quickly visualizing the mathematical modeling of physical **flux sources** or **sinks**.

Now, let us review a few, simple 2-D flux diagrams. Since $f(z) = f(r\operatorname{cis}\theta) = U + iW = \ln(z)$, with $U(r,\theta) = \ln(r)$ and $W(r,\theta) = \theta$, $-\pi < \theta \leq \pi$ (thus $\ln(z)$ has a branch cut at $\theta = \pi$), then the streamlines $W = $ constant are **rays** emanating from the origin, and the equipotential lines $U = $ constant, are circumferences centered at the origin.

The mathematical singularity at the origin, exhibited by $\ln(z)$, corresponds to the **physical singularity**, which, in this case, is due to a concentrated source of flux at that point (as well as "a sink at infinity"). The function $\ln(\frac{1}{z}) = -\ln(z)$ corresponds to a sink of flux at the origin (and an associated "source at infinity").

As shown above, the magnitude Q of the outward flux across C (emanating from the origin), is given by

$$Q = \operatorname{Im} \int_C \frac{1}{z}\, dz = 2\pi. \tag{2.58}$$

Analogously, the magnitude of the inward flux across C into the sink is (-2π). Therefore, the function $\frac{1}{2\pi}\ln(z)$ is the complex potential, describing the source of the unit flux at 0.

Successive complex derivatives of $f(z) = \ln(z)$ also have corresponding physical counterparts, which are the **multipoles** placed at the origin.

In Figures 2.2 and 2.3, the dotted lines indicate **branch cuts** which exhibit a sequence of increasing number of singularity distributions over the complex z-plane.

Notice that, in the case of a coupled source and sink of equal strength, measured by a 2π exchange flux, and distant $(2h)$ from each other:

$$\lim_{h\to 0} \frac{f(z)}{2h} = -\lim_{h\to 0} \left(\frac{1}{2h}\right)\ln\left[\frac{(z+h)}{(z-h)}\right] = -\frac{1}{z}. \tag{2.59}$$

Let M be a positive constant, and consider a coupled source and sink with a $2\pi C(h)$ exchange flux, with $C(h)$ to be defined presently.

The one-parameter family of complex potentials,

$$F_h(z) = C(h)\ln(z-h) - C(h)\ln(z+h), \tag{2.60}$$

which, as a result of h being a real and positive value, has the following properties:

$$\lim_{h\to 0} F_h(z) = -\lim_{h\to 0}\left(\frac{\ln(z+h) - \ln(z-h)}{2h}\right)\lim_{h\to 0}(2hC(h))). \tag{2.61}$$

Now, choose $C(h)$, such that

$$\lim_{h\to 0}(2hC(h)) = M. \tag{2.62}$$

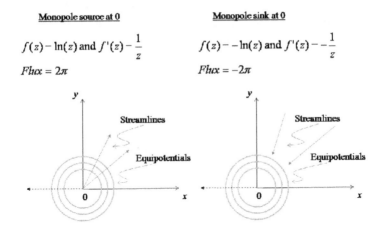

Figure 2.2: Monopoles.

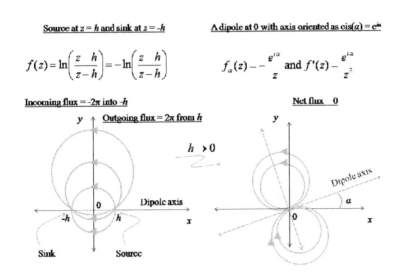

Figure 2.3: Dipoles.

In particular, with $C(h) = \frac{M}{2h}$:

$$\lim_{h \to 0} F_h(z) = -M\left(\frac{1}{z}\right). \tag{2.63}$$

Since $M = C(h)(2h)$, and $C(h)$ measures the strength of the outgoing source flux (except for the factor 2π), then it is only natural to denote M as the **dipole moment** associated with the complex potential $f(z) = -M\left(\frac{1}{z}\right)$, which then models an ideal physical dipole located at 0, with the real axis being a dipole axis and dipole moment M (real physical dipoles have the coupled source and sink at some small but **finite** distance apart).

It is now clear that

$$f_\alpha(z) = -\frac{e^{(i\alpha)}}{z} \tag{2.64}$$

is the complex potential modeling a physical dipole with an axis rotated by α with respect to the positive real axis in the z-plane.

It then becomes apparent that the streamlines and equipotentials associated with the complex potential

$$f(z) = -\frac{M}{z} = U(x,y) + iW(x,y) \tag{2.65}$$

are orthogonal families of circumferences with centers laid along the real and imaginary axes:

$$f(z) = -\frac{M\bar{z}}{r^2} = -\frac{Mx}{x^2+y^2} + i\frac{My}{x^2+y^2}, \tag{2.66}$$

so that

$$U(x,y) = -\frac{Mx}{x^2+y^2}, \tag{2.67}$$

and

$$W(x,y) = \frac{My}{x^2+y^2}. \tag{2.68}$$

When the multipoles are located at a point γ other than the origin, then the corresponding complex potential is simply $f(z-\gamma)$.

It should be noted that many interesting potential flows can be better understood through the composition of complex variable functions, with a means of describing 2-D flows with boundaries restricting the flow to be tangential to these lines in the complex plane.

For example, the complex potential given by $f(z) = z$ with a corresponding velocity field $\overline{f'(z)} = 1$ models a uniform flow in the x direction of the z-plane, and also has a unit magnitude. Therefore, when you look at the complex potential $F(\omega) = \omega$ in the complex ω-plane, you can see a uniform flow in the direction of the positive real axis. Now, consider a composition of functions $F(\omega) = F(g(z))$ as a complex potential for the z-plane, with $\omega = g(z) = z^2$, so that $F(g(z)) = z^2$. We now see that the real axis in the ω-plane is the image of the rectangular boundary in the z-plane, which is made up of the positive real axis and imaginary semiaxis, and the streamlines are described by the quantity $2xy$ being constant. The tangential condition for the velocity field $\overline{F'(z)}$ states that the normal derivative of the real part U vanishes along the fluid boundaries, and is called an homogeneous **Neumann boundary condition**,

$$\frac{dU}{dn} = 0, \tag{2.69}$$

which supplements Laplace's equation for U.

In the 2-D case, a consequence of the above-mentioned physical considerations, is that the singularity points of the continuum fields, as well as the boundary points of the corresponding spatial domains, must be thought of as a model for locations where singular physical phenomena take place, or, for interfaces where changes in the nature of extended substances occurs. Keeping this in mind may be of help while trying to solve various problems: both analytically in the continuous version, or numerically in the discrete version. This is especially true regarding the formulation of adequate boundary conditions to be satisfied by the field under consideration.

Flux lines emanate from a source location and always end at a sink location, which may be infinitely distant from the original source, as in the case of the complex potential $f(z) = \ln(z)$.

2.2   General 3-D Use of Flux Vector Densities

As shown in Appendix I, if, instead of looking at vector fields such as the velocity field $\mathbf{v}$ or $\rho\mathbf{v}$, we consider a general vector field $\mathbf{j}$, with flux lines crossing a curved and oriented smooth surface σ from its w-side to the e-side (think of west and east as opposite directions on a map), then the associated flux across σ from the w-side to the e-side can be defined as

$$\Phi_{we}(\sigma) = \int_{\sigma} \mathbf{j} \cdot \mathbf{n}_{we} \, d\sigma. \tag{2.70}$$

(The field of normals must be indexed, since there is no outward sense defined for $\mathbf{n}$ when σ is not a closed surface.)

2.2.1   Important Particular Cases

1. In the case of $\mathbf{j} = \left(\frac{1}{\rho}\right)\mathbf{E}$, where ρ is the resistivity of an electrical conductor acted upon by an electric field $\mathbf{E}$, such that the electric current intensity I flowing through some cross section σ of the conductor is given as

$$I(\sigma) = \int_{\sigma} \left(\frac{1}{\rho}\right) \mathbf{E} \cdot \mathbf{n} \, d\sigma. \tag{2.71}$$

For constants $\mathbf{E}$ and ρ, this leads to the familiar **Ohm's law**: $I = V/R$, where V is the voltage difference giving rise to $\mathbf{E}$, and R is the total ohmic resistance between the conductor's end points.

2. In the case of $\mathbf{j} = \mathbf{B}$, the magnetic induction field, so that the **magnetic flux Φ** is

$$\Phi(\sigma) = \int_\sigma \mathbf{B} \cdot \mathbf{n} \, d\sigma. \tag{2.72}$$

This is used to express **Faraday's law**.

3. For the case of $\mathbf{j} = \mathbf{D}$, the electric displacement field, and the **electric flux** is given by

$$\Phi(\sigma) = \int_\sigma \mathbf{D} \cdot \mathbf{n} \, d\sigma. \tag{2.73}$$

This is used to express **Gauss' law**.

4. For $\mathbf{j} = -k \, grad \, T$ (**Fourier's law**), where k is the heat conductivity, and T is the scalar temperature field, such that the heat flow is given as

$$\Phi_{we}(\sigma) = -\int_\sigma k \, grad \, T \cdot \mathbf{n}_{we} \, d\sigma. \tag{2.74}$$

5. When $\mathbf{j} = -k \, grad \, T$ (**Fick's law**), where k is the diffusivity of a medium in which some substance with concentration C can be diffused. The diffusive flow is given by

$$\Phi_{we}(\sigma) = -\int_\sigma k \, grad \, C \cdot \mathbf{n}_{we} \, d\sigma. \tag{2.75}$$

This list is by no means exhaustive, but does give an idea of the many applications for which the flux notion is needed in order to formulate mathematical laws in the form of equations that model physical laws (and which usually express some instance of a conservation principle). Examples of these types of equations are included in (§2.3.1).

2.3   Illustrative Examples of PDEs

2.3.1   The Continuity Equation

Let $M(t)$ be the mass inside some region Ω at time t, which has accumulated there as a result of being transported by a fluid with mass density $\rho(\mathbf{x}, t)$, and velocity field $\mathbf{v}(\mathbf{x}, t)$, such that

$$M(t) = \int_\Omega \rho(\mathbf{x}, t)\, dV. \tag{2.76}$$

Suppose also that there are mass sources inside Ω which are described by some scalar field $S(\mathbf{x}, t)$, so

$$\int_\Omega S(\mathbf{x}, t)\, dV = \dot{M}_S(t). \tag{2.77}$$

Here, $\dot{M}_S(t)$ stands for the time rate of change for the mass M_S as generated by the distributed sources.

Now, the net outflux of mass through σ, the closed surface boundary of Ω, is given by

$$\int_\Omega \mathbf{n}(\mathbf{x}) \cdot (\rho \mathbf{v})(\mathbf{x}, t)\, d\sigma. \tag{2.78}$$

If the net mass outflux is positive, then the mass $M_\mathbf{v}$ inside Ω must be decreasing per unit (differential) time as a result of the velocity flux density $\mathbf{v}$, so that

$$\dot{M}_\mathbf{v} = -\int_\sigma \mathbf{n}(\mathbf{x}) \cdot (\rho \mathbf{v})(\mathbf{x}, t)\, d\sigma = -\int_\Omega \nabla \cdot (\rho \mathbf{v})(\mathbf{x}, t)\, dV, \tag{2.79}$$

for differentiable $\mathbf{j} = \rho \mathbf{v}$. Since $M(t) = M_S(t) + M_\mathbf{v}(t)$, then

$$\dot{M}_s(t) + \dot{M}_\mathbf{v}(t) = \int_\Omega S(\mathbf{x}, t)\, dV - \int_\Omega \nabla \cdot (\rho \mathbf{v})(\mathbf{x}, t)\, dV = \frac{d}{dt} \int_\Omega \rho(\mathbf{x}, t)\, dV. \tag{2.80}$$

If the region Ω, also called the **control volume** in fluid dynamics, does not change with time t, then

$$\int_\Omega \left(\frac{\partial \rho}{\partial t} + \nabla \cdot (\rho \mathbf{v}) - S \right)(\mathbf{x}, t)\, dV = 0. \tag{2.81}$$

The arbitrariness of Ω implies that, for a continuous integrand

$$\frac{\partial \rho}{\partial t} + \nabla \cdot (\rho \mathbf{v}) = S, \tag{2.82}$$

for all $\mathbf{x} \in \Omega$ and $t \in I$, where I is some real interval.

In the absence of mass sources (or sinks) inside Ω, we get

$$\frac{\partial \rho}{\partial t}(\mathbf{x}, t) + \nabla \cdot (\rho \mathbf{v})(\mathbf{x}, t) = 0. \tag{2.83}$$

This is the "continuity equation" or **mass conservation law**, since when $\nabla \cdot (\rho \mathbf{v}) < 0$, then

$$-\int_{\Omega} \nabla \cdot (\rho \mathbf{v})(\mathbf{x}, t) \, dV = -\int_{\sigma} \rho \mathbf{v} \cdot \mathbf{n}(\mathbf{x}, t) \, d\sigma > 0, \qquad (2.84)$$

that is, **the net mass outflux** is less than zero, so that influx $>$ outflux inside Ω, which results in $\frac{\partial \rho}{\partial t} > 0$. Since there is no creation or destruction of mass inside Ω, the equality $\frac{\partial \rho}{\partial t} = -\nabla \cdot (\rho \mathbf{v})$ reflects the fact that, per unit (differential) time inside the region with differential volume dV and an associated mass $\rho \, dV$, then

$$\frac{\partial \rho}{\partial t} = \text{accumulated mass} = \text{incoming mass flux} - \text{outgoing mass flux}, \quad (2.85)$$

stating that mass inside Ω was neither created nor destroyed.

Also, if a fluid is incompressible (i.e., its density is constant), then the equation of continuity implies that

$$\nabla \cdot \mathbf{v} = 0 \qquad (2.86)$$

in Ω.

2.3.2   Maxwell's Equations

Faraday's law states that the electromotive force induced along a closed curve C bounding a surface σ is equal to the negative of the time rate of change of the magnetic flux through σ:

$$\int_{C} \mathbf{E} \cdot d\mathbf{r} = -\frac{d}{dt} \int_{\sigma} \mathbf{B} \cdot \mathbf{n} \, d\sigma = -\int_{\sigma} \frac{\partial \mathbf{B}}{\partial t} \cdot \mathbf{n} \, d\sigma, \qquad (2.87)$$

and by the Stokes' theorem,

$$\int_{C} \mathbf{E} \cdot d\mathbf{r} = \int_{\sigma} curl \, \mathbf{E} \cdot \mathbf{n} \, d\sigma, \qquad (2.88)$$

so that

$$\int_{\sigma} curl \, \mathbf{E} \cdot \mathbf{n} \, d\sigma = \int_{\sigma} \left(-\frac{\partial \mathbf{B}}{\partial t} \right) \cdot \mathbf{n} \, d\sigma. \qquad (2.89)$$

In the case of continuously differentiable fields, the arbitrariness of σ implies **Maxwell's first equation**:

$$curl \, \mathbf{E} = -\frac{\partial \mathbf{B}}{\partial t}. \qquad (2.90)$$

Now, for the electric displacement field $\mathbf{D} = \epsilon\mathbf{E}$, if ρ stands for the volumetric charge density, then

$$\int_{\sigma} \mathbf{D} \cdot \mathbf{n} \, d\sigma = \int_{\Omega} \rho dV. \tag{2.91}$$

Using Gauss' divergence theorem, we see that

$$\int_{\Omega} div \, \mathbf{D} \, dV = \int_{\Omega} \rho \, dV. \tag{2.92}$$

In this case, the arbitrariness of Ω implies **Maxwell's second equation**:

$$div \, \mathbf{D} = \rho. \tag{2.93}$$

Since magnetic monopoles are yet to be found in nature, no field sources are available for $\mathbf{B}$, which leads to **Maxwell's third equation** throughout Ω:

$$div \, \mathbf{B} = 0. \tag{2.94}$$

Letting the magnetic field $\mathbf{H} = \left(\frac{1}{\mu}\right)\mathbf{B}$, where μ stands for the magnetic permeability, **Maxwell's fourth equation** states that

$$curl \, \mathbf{H} = k\mathbf{E} + \frac{\partial \mathbf{D}}{\partial t}, \tag{2.95}$$

where k stands for the electric conductivity.

In a vacuum, $k = \rho = 0$ in Ω, so that the system M_x of Maxwell's equations becomes homogeneous, such that

$$(M_x) \begin{cases} curl \, \mathbf{E} = -\mu\frac{\partial \mathbf{H}}{\partial t} \\ div \, \mathbf{E} = 0 \\ div \, \mathbf{H} = 0 \\ curl \, \mathbf{H} = \epsilon\frac{\partial \mathbf{E}}{\partial t} \end{cases} \tag{2.96}$$

An easy manipulation of M_x shows that, in a vacuum, both $\mathbf{E}$ and $\mathbf{H}$ satisfy the following vector-wave equation for an unknown $\mathbf{u}(\mathbf{x}, t)$:

$$\left(grad \, div - curl \, curl - \left(\frac{1}{c^2}\right)\frac{\partial^2}{\partial t^2}\right)\mathbf{u} = 0. \tag{2.97}$$

Here, $c^2 = \frac{1}{\epsilon\mu}$.

This shows that, in a vacuum, the electromagnetic field $(\mathbf{E}, \mathbf{H})$ propagates as vector waves with a speed c, and $\epsilon = \epsilon_0$, $\mu = \mu_0$ in a vacuum.

The second-order operator ($grad \, div$ - $curl \, curl$) is also known as the **vector Laplacian**, due to the fact that, in Cartesian coordinates, when $\mathbf{u} = u_1\mathbf{i} + u_2\mathbf{j} + u_3\mathbf{k}$, then

$$(grad \, div - curl \, curl)\mathbf{u} = \Delta u_1\mathbf{i} + \Delta u_2\mathbf{j} + \Delta u_3\mathbf{k}. \tag{2.98}$$

Some authors define the vector Laplacian as only $curl \, curl$, while others also consider the "Hodge Laplacian" to be defined as ($curl \, curl$ - $grad \, div$).

2.4   A Comment on the Numerical Treatment of the *grad* Operator

With regard to the numerical treatment of the *grad* operator, it is important to understand what physical role is being played by *grad f* in a particular model, since there are three different situations which may arise. These situations include the following:

1. When $f = U$, and $grad\, U = \mathbf{E}$ (the electric field force per unit charge), then $grad\, U$ acts as a force field along a 1-D trajectory, followed up by a charge q under the force $q\mathbf{E}$. In this case, $grad\, f$ should be numerically evaluated at the centers of the line segments along the trajectory.

2. Now, if $f = T$ or C (temperature or concentration), then $grad\, f$ acts as a **flux** vector density acting upon a 2-D surface σ, and $grad\, f$ should be numerically evaluated upon the faces of the 3-D cells. This situation usually leads to the consideration of *div grad f* while modeling a physical conservation law (thermal energy or solute mass conservation for T or C, respectively).

3. If $f = p$ (pressure field) throughout a static fluid with a density ρ, under gravity force $(-\rho g\mathbf{k})$ per unit volume, then $\left(\frac{1}{\rho} grad\, p = -g\mathbf{k} = \mathbf{b} \right)$ acts as a body force $\mathbf{b}$ per unit mass throughout a 3-D region Ω, so that when the boundary of Ω is a closed piecewise smooth surface σ, and the total force $\mathbf{W}$ acting upon the fluid at rest occupying Ω is given by

$$\mathbf{W} = \int_\Omega \rho\mathbf{b}\, dV = \int_\Omega grad\, p\, dV = \int_\sigma p\mathbf{n}\, d\sigma. \qquad (2.99)$$

As can be seen in Appendix C, this is **Archimedes' buoyancy law** for $\mathbf{b} = -g\mathbf{k}$, since, in the case of a submerged, nonporous solid body which comes to occupy Ω inside the static fluid, $\mathbf{W}$ is merely the weight of the fluid displaced from Ω after the submersion of the body in it. In this case, p should be numerically evaluated upon the faces of the 3-D cells, while $grad\, p$ should be numerically evaluated at the centers of 3-D cells.

2.5   Concluding Remarks

These examples demonstrate the many reasons why numerical methods are so effective in solving a large number of important problems in science and engineering. Also, recall that these methods use discrete operators which model the fundamental first-order operators *div*, *grad*, and *curl*, and their compositions, in a way that **mimics** a grid function's basic differential properties (including *curl grad* $\equiv 0$, and *div curl* $\equiv 0$). They should also mimic the continuum conservation principle, which has been modeled in integral form by the extended version of Gauss' divergence theorem (2.9) and all of its vector-valued related forms, and is listed as Formula (1.1) on page 41 of [65].

When utilizing Castillo–Grone's discrete operators, the spatially related portion of the PDE under examination **is not** discretized term-by-term, as is done with standard finite difference methods, but it is discretized considering only the terms which can be modeled as a result of applying mimetic discrete **operators** and their compositions. Also, by using staggered grids, along with generalized inner products for the finite-dimensional vectors approximating the continuous exact solution, it is possible to obtain any even order accurate discrete solutions **uniformly up to the boundary** of the physical domain, as illustrated in Chapter 4.

2.6 Sample Problems

1. Consider an arbitrary surface:

$$z(x,y) = 1 - x^2 - y^2 \qquad (2.100)$$

and let σ be the portion that lies above the xy-plane. Suppose that σ is oriented up as in Figure 2.4.

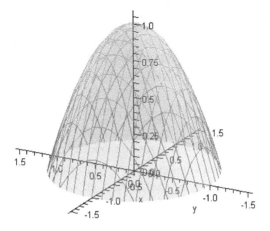

Figure 2.4: Figure for Problem 1. Surface $z(x,y) = 1 - x^2 - y^2$.

Find the flux of the vector field $\mathbf{F}(x, y, z) = x\mathbf{i} + y\mathbf{j} + z\mathbf{k}$ across σ.

2. Consider the following instance of the continuity equation (2.83) for the concentration c of a certain chemical species:

$$\frac{\partial c}{\partial t} = -\frac{\partial \mathbf{v}}{\partial x} \pm S. \qquad (2.101)$$

Consider a 1-D capillary vessel Ω of area $A(x, t)$.

(a) What is the meaning of

$$\int_\Omega c(x, t) A(x, t) dx? \qquad (2.102)$$

(b) What is the meaning of

$$\int_\Omega S(x, t) A(x, t) dx? \qquad (2.103)$$

(c) Deduce the following:

$$\frac{\partial}{\partial t} c(x_0, t) A(x_0, t) = -\frac{\partial}{\partial t} [\mathbf{v}(x_0, t) A(x_0, t)] \pm S(x_0, t) A(x_0, t). \qquad (2.104)$$

3. Consider a 1-D thin rod Ω which has no heat sources nor sinks and which has a non-insulated lateral surface area.

(a) Let $w(x, t)$ denote the heat energy flowing out the lateral sides per unit surface area. Derive the PDE for the temperature $u(x, t)$.

(b) Let $\gamma(x, t)$ denote the temperature of the environment. Let P and A denote the rod perimeter and cross-sectional area, respectively. Assume $w(x, t)$ is proportional to the temperature difference between the rod $u(x, t)$ and $\gamma(x, t)$. Derive the following equation:

$$c\rho \frac{\partial u}{\partial t} = \frac{\partial}{\partial x}\left(k_0 \frac{\partial u}{\partial x}\right) - \frac{P}{A}[u(x, t) - \gamma(x, t)] h(x), \qquad (2.105)$$

where $h(t)$ is a positive proportionality coefficient.

(c) Adapt the latter equation to a rod of circular cross-section with constant thermal properties and an outside temperature of zero degrees.

4. Consider:

$$\frac{\partial u}{\partial t} = k\nabla^2 u, \qquad (2.106)$$

for $t > 0$ and $0 < x < L$. Consider the following boundary conditions (also for $t > 0$):

$$u(0, t) = 0 \qquad (2.107)$$
$$u(L, t) = 0 \qquad (2.108)$$

and consider the following initial condition:

$$u(x, 0) = f(x). \qquad (2.109)$$

(a) Analytically solve the proposed PDE. Is there any equilibrium solution $u_e(x)$?

(b) Describe and comment upon any ODEs attained while solving the proposed PDE.

(c) Let $f(x) = 100$. Produce a computer plot of the analytical solution. Is there any computational constraint arising while attempting this? Can you physically explain the resulting plot? Repeat the problem but using $f(x) = u_e(x)$ instead, where $u_e(x)$ denotes the equilibrium solution you should have found in Part 1.

5. According to E. M. von Hornbostel and C. Sachs in their article entitled *Systematik der Musikinstrumente* [141], guitars can be classified as **chordophones**; a type of musical instrument where sound is produced by one or more vibrating strings. It is clear that the evolution of the guitar as a musical instrument has been tremendous, and so is the evolution of the related technology, such as electronic and stroboscopic tuners. A strobe tuner shows the difference between a reference frequency and the musical note and the latter is based on the vibrations of the string. We are interested in the mathematical modeling of a vibrating string. In order to analyze the behavior of a string held at both ends, we will focus our attention to the displacement y as a function of both time t and position x along the string. Let T denote the tension of the string and ρ its linear density. The equation describing the vibration of the string is called the **one-dimensional wave equation**:

$$\frac{\partial^2 y}{\partial t^2} = c^2 \frac{\partial^2 y}{\partial x^2}, \qquad (2.110)$$

where $c = \sqrt{T/\rho}$.

(a) Define the following differential operator:

$$\frac{\partial^2}{\partial t^2} - c^2 \frac{\partial^2}{\partial x^2} \qquad (2.111)$$

Can it be factored in a useful way?

(b) Considering $u = x + ct$, $v = x - ct$ and the multivariable form of the chain rule, deduce the following expressions:

$$\frac{\partial^2 y}{\partial t^2} = c^2 \left(\frac{\partial^2 y}{\partial u^2} - 2\frac{\partial^2 y}{\partial u \partial v} + \frac{\partial^2 y}{\partial v^2} \right), \qquad (2.112)$$

$$\frac{\partial^2 y}{\partial x^2} = \frac{\partial^2 y}{\partial u^2} - 2\frac{\partial^2 y}{\partial u \partial v} + \frac{\partial^2 y}{\partial v^2}. \qquad (2.113)$$

(c) Finally, justify rewriting Equation (2.110) as

$$\frac{\partial^2 y}{\partial u \partial v} = 0, \qquad (2.114)$$

integrate and show that, for suitable chosen functions f and g, the general solution of the wave equation (2.110) is given by

$$y = f(x + ct) + g(x - ct). \qquad (2.115)$$

Chapter 3

Notes on Numerical Analysis

In this chapter, we present some of the basic numerical analysis techniques that are especially helpful when utilizing this book for a course on the numerical solutions of partial differential equations.

3.1 Computational Errors

The study of the methods used for finding approximate solutions to partial differential equations includes the analysis of the resulting errors, as well as the bounds on these errors. There are basically three types of errors: The first comes from the data and is usually a result of physical measurement methods' inherent limitations. The other two errors, called the roundoff error and the discretization or truncation error, respectively, are a result of the actual numerical calculations.

Roundoff errors arise when computations are performed in finite precision on a digital computer (i.e., the computations are based on quantities whose representation is restricted to a finite number of digits). As a result, we can define the **roundoff error** as the amount that the computational solution differs from a solution obtained when exact arithmetic is used.

Generally speaking, a numerical method is only designed to approximate the exact solution of a given partial differentiation equation; the type of discretization yields the discretization or truncation error. We define the **truncation error** as the difference between the projection to a grid point of the derivative of a smooth function, and the discrete difference approximation of the derivative using values of the smooth function projected to the grid points. Below, the **cell projection** operator p_h maps a smooth scalar function to discrete cell-valued functions:

$$(p_h\, u)_{i+1/2} = \tilde{u}_{i+1/2} \equiv u(\tilde{x}_{i+1/2}). \qquad (3.1)$$

Here, the **nodal projection** operator, P_h, maps a smooth vector function to its values at the nodes:

$$(P_h\, \mathbf{w})_i = \tilde{w}_i \equiv \mathbf{w}(\tilde{x}_i). \qquad (3.2)$$

If **w** is a smooth vector function, then the truncation error of the discrete divergence ψ_{div} is the nodal function

$$\psi_{div}\mathbf{w} = p_h(\frac{d\mathbf{w}}{dx}) - div(P_h\mathbf{w}). \tag{3.3}$$

However, if u is a smooth scalar function, then the truncation error of the discrete gradient ψ_{grad} is

$$\psi_{grad}u = grad(p_h u) - P_h(\frac{du}{dx}). \tag{3.4}$$

3.2 Order of Accuracy

We define the discrete inner products of functions defined on a staggered grid by generalizing the usual inner products, using the matrices Q and P:

$$\langle Q, f, g \rangle \triangleq \sum_{i,j=0}^{N-1} Q_{i+\frac{1}{2}, j} + \frac{1}{2} f_{i+\frac{1}{2}} g_{j+\frac{1}{2}} \tag{3.5}$$

and

$$\langle u, Pv \rangle \triangleq \sum_{i,j=0}^{N} P_{i,j} u_i v_j. \tag{3.6}$$

For these expressions to be inner products, P and Q must be symmetric positive-definite operators.

The general divergence and gradient are defined by

$$(div\ \mathbf{v})_{i+\frac{1}{2}} \triangleq \frac{1}{dx} \sum_{j=0}^{N} d_{i+\frac{1}{2}, j} v_j, \quad 0 \le i \le N-1 \tag{3.7}$$

$$(grad\ f)_i \triangleq \frac{1}{dx} \left(g_{i,0} f_0 + \sum_{j=0}^{N-1} g_{i,j+\frac{1}{2}} f_{j+\frac{1}{2}} + g_{i,N} f_N \right), \quad 0 \le i \le N. \tag{3.8}$$

and satisfy a discrete analog of the extended Gauss divergence theorem, when

$$\langle Qdiv\ \mathbf{v}, f \rangle + \langle \mathbf{v}, Pgrad\ f \rangle = v_N f_N - v_0 f_0, \tag{3.9}$$

is satisfied.

The operators have an **order of accuracy** k, when

$$(div\ \mathbf{v})_{i+\frac{1}{2}} - v'((i+\frac{1}{2})h) = O(dx^k), \quad 0 \le i \le N-1 \tag{3.10}$$

$$(grad\ f)_i - f'(ih) = O(dx^k), \quad 0 \le i \le N. \tag{3.11}$$

Our goal is to define Q, P, div, and $grad$, thereby satisfying the extended Gauss divergence theorem (see Appendix A), as well as having the same order of accuracy in the interior of the domain as well as at the boundary. For some PDEs, the reduced accuracy near the boundary has a small effect on the accuracy of the solution; however, in many problems, having the same order of accuracy improves the quality of the solution.

The leading error term in approximations of the first derivative by finite differences is

$$C, dx^k, \frac{d^{k+1} f}{dx^{k+1}}, \tag{3.12}$$

where C is a constant. Therefore, an equivalent formulation of the accuracy requirement must have the derivatives that exact on the polynomials 1, x, ..., x^k:

$$\left(div \ x^j\right)_{i+\frac{1}{2}} - j\left((i + \frac{1}{2})h\right)^{j-1} = 0, \quad 0 \le i \le N - 1, \tag{3.13}$$

$$\left(grad \ x^j\right)_i - j, (i, h)^{j-1} = 0, \quad 0 \le i \le N, \tag{3.14}$$

for $0 \le j \le k$. Finally, we say that a discrete differential operator is **k-th order accurate** when it yields an exact answer when applied to polynomials of degree k or less.

When estimating errors in numerical solutions, it is important to understand that when errors are expected to behave like some power of h, that is, if the error $E(h)$ behaves like $E(h) \approx Ch^p$, then $\log |E(h)| \approx \log |C| + p \log h$, so that, in a log-log scale, the error behaves linearly with a slope that is equal to p, which is the order of accuracy as usually defined in finite difference methods. A first step to test a code and ensure that it is producing correct results with the expected accuracy is to try it on a problem for which the exact solution is known, in which case we can compute the error in the numerical solution exactly. Besides checking that the error is small in some grid, we can also refine the grid and check how the error is behaving asymptotically as $h \to 0$, to verify that the expected order of accuracy p, and perhaps even the error constant C, is seen on a log-log scale.

3.3   Norms and Condition Numbers

In order to measure the size of vectors and matrices (and, particularly, of the errors involving them), we need to provide our linear space with a norm. We define the **vector norm** $|| \cdot || : \mathbb{R}^n \mapsto \mathbb{R}^+ = \{x \in \mathbb{R} : x > 0\}$ as a real-valued function for which

1. $||\mathbf{x}|| = 0 \Leftrightarrow \mathbf{x} = 0$,

2. $||\gamma\mathbf{x}|| = |\gamma|||\mathbf{x}||$, for all $\gamma \in \mathbb{R}$,

3. $||\mathbf{x} + \mathbf{y}|| \leq ||\mathbf{x}|| + ||\mathbf{y}||$ (triangular inequality).

Now, we define the **Holder norms** as

$$||\mathbf{x}||_p \triangleq \left(\sum_{i=1}^{n} |x_i|^p\right)^{1/p}, \; p \geq 1. \tag{3.15}$$

Note that, for $p = 1$, we obtain the **linear norm**, while for $p = 2$ we obtain the **quadratic** or **Euclidean norm**, and for $p = \infty$ we obtain the **maximum** or **infinity norm**:

$$||\mathbf{x}||_\infty \triangleq \max_{1 \leq i \leq n} |x_i|. \tag{3.16}$$

Since they are equivalent, all norms on a finite dimensional space define the same topology. Nevertheless, for some applications, one particular norm may be preferred over another.

We can also define a norm for matrices. A **matrix norm** $||\cdot|| : \mathbb{R}^{n \times n} \mapsto \mathbb{R}^+$ is a real-valued function, for which

1. $||A|| = 0 \Leftrightarrow A = 0$,

2. $||\gamma A|| = |\gamma|||A||$, for all $\gamma \in \mathbb{R}$,

3. $||A + B|| \leq ||A|| + ||B||$.

Furthermore, if

$$||AB|| \leq ||A||||B||, \tag{3.17}$$

then it is called a **multiplicative** norm.

In this text, we mainly deal with matrix norms derived from **associated least upper bounds** for vector norms,

$$||A||_p \triangleq \max_{\mathbf{x} \neq 0} \frac{||A\mathbf{x}||_p}{||\mathbf{x}||_p}. \tag{3.18}$$

Analogously, we define the **greatest lower bound** as

$$||A||_p \triangleq \min_{\mathbf{x} \neq 0} \frac{||A\mathbf{x}||_p}{||\mathbf{x}||_p}. \tag{3.19}$$

We are able to deduce some important results from the latter definitions: Analogous to the previously presented Holder norms for vectors, we have, for different values of p:

$$||A||_1 = \max_j \sum_i |a_{ij}|, \tag{3.20}$$

$$||A||_2 = (\rho(A^H A))^{1/2}, \tag{3.21}$$

$$||A||_\infty = \max_i \sum_j |a_{ij}|, \tag{3.22}$$

where, $\rho(A)$ is defined as the **spectral radius** of the matrix A.

Another important result is that acquired when Q_1 and Q_2 are arbitrary orthogonal matrices; in this case $||Q_1 A Q_2||_2 = ||A||_2$. The result indicates why orthogonal matrices are so important in numerical analysis: In many applications, we are required to multiply a certain matrix by many transformations matrices. If they are orthogonal, and we have a method to compute them reasonably well, the resulting product will have roughly the same 2-norm; i.e., it will be **numerically stable**.

3.3.1 Condition Number of a Matrix

While analyzing theoretical aspects of the numerical solutions of PDEs, including their solvability by means of the singularity of the involved matrix defining the arising system of linear equations, an important characteristic is the condition number of this matrix. For a given nonsingular matrix A, we define and denote its **condition number** as

$$\text{cond}_k(A) = ||A||\,||A^{-1}||. \tag{3.23}$$

We note that $\text{cond}_k(A) \geq 1$ and $\text{cond}_k(\gamma A) = \text{cond}_k(A)$ for any $\gamma \in \mathbb{R}$.

The condition number of a nonsingular matrix A is directly related to its singular values. First, recall that $||A||_2 = \left(\rho(A^H A)\right)^{1/2} = \sigma_1$. If we consider a singular-value decomposition, i.e., $A = Q_1^H \Sigma Q_2$, where $\Sigma = diag(\sigma_1, ..., \sigma_n)$, then $||A^{-1}||_2 = ||\Sigma^{-1}||_2 = \sigma_n^{-1}$. Therefore:

$$\text{cond}_2(A) = \frac{\sigma_1}{\sigma_n}. \tag{3.24}$$

We can now relate this result to the singularity of the matrix by noting that, as σ_n tends to zero (i.e., the matrix becomes more numerically singular), the condition number gets larger; therefore, a large condition number implies complications when it comes to the solvability of the problem. A matrix with a large condition number is said to be **ill-conditioned**, and so is said to be the related problem the matrix is helping to model.

3.4 Linear Systems of Equations

Systems of equations arise within many problems in science and engineering. Particularly, when numerically solving a linear PDE, we end up solving a system of linear equations $A\mathbf{x} = \mathbf{b}$. Direct solvers can be used if we can afford to factor the matrix, but it is often necessary to utilize iterative methods instead. If the partial differential equation is nonlinear, then a nonlinear system of equations must be solved. The rest of this chapter is devoted to the introduction of the numerical methods intended to solve the resulting system of equations. Specifically, in this section we will focus on solving linear systems of equations.

3.4.1 Direct Methods

First, we focus on direct methods, in which the solution is found after a predetermined or well-defined number of operations. Some well-known direct methods include (but are not limited to):

1. Gaussian elimination, with backward substitution, partial pivoting, and scaled partial pivoting

2. The Thomas algorithm for tridiagonal systems

3. LU factorization

4. QR factorization

5. Singular-values decomposition

6. Cholesky factorization

These methods are explained in greater detail in the bibliography. Given the subtleties among different programming languages, we must first distinguish between row-oriented and column-oriented algorithms. Similarly, we must also differentiate between two types of direct methods: Methods which act upon dense linear systems, and methods acting on sparse linear systems. A well-known example of a direct method for dense linear systems is **LU factorization**. When we use this method, to solve $A\mathbf{x} = \mathbf{b}$ (after first factoring A into the product LU), we obtain

$$A\mathbf{x} = \mathbf{b} \Rightarrow LU\mathbf{x} = \mathbf{b} \Rightarrow L\mathbf{y} = \mathbf{b}. \tag{3.25}$$

Now, given its properties, the system $L\mathbf{y} = \mathbf{b}$ is easy to solve using **forward substitution**:

$$y_i = b_i - \sum_{j=1}^{i-1} l_{ij} y_j. \tag{3.26}$$

Once we know the solution for the latter system, we can then solve $U\mathbf{x} = \mathbf{y}$ by means of **backward substitution**, so that

$$x_i = \frac{1}{u_{ii}} \left[y_i - \sum_{j=i+1}^{m} u_{ij} x_j \right]. \tag{3.27}$$

A standard algorithm for solving sparse linear systems is the **Thomas algorithm for tridiagonal systems**, which is a simplified form of Gaussian elimination, and is used to solve tridiagonal systems of equations of the form $a_i x_{i-1} + b_i x_i + c_i x_{i+1} = d_i$. This algorithm exploits the fact that the matrix has a tridiagonal structure, thereby allowing the computation of the solution in $O(n)$ operations instead of $O(n^3)$ as required by Gaussian elimination. With this method, the solutions are given by

$$x_n = d'_n, \tag{3.28}$$
$$x_i = d'_i - c'_i x_{i+1}, \tag{3.29}$$

where

$$c'_i = \begin{cases} \frac{c_i}{b_i}; & i = 1 \\ \frac{c_i}{b_i - c'_{i-1} a_i}; & i = 2, 3, ..., n-1, \end{cases} \tag{3.30}$$

and

$$d'_i = \begin{cases} \frac{d_i}{b_i}; & i = 1 \\ \frac{d_i - d'_{i-1} a_i}{b_i - c'_{i-1} a_i}; & i = 2, 3, ..., n. \end{cases} \tag{3.31}$$

3.4.2 Iterative Methods

In contrast to direct methods, iterative methods involve the repetition of a specific procedure, with the solution approaching the exact one during these iterations. These methods generally require stopping criteria, which are usually defined in terms of the desired accuracy of the computed solution.

Well-known iterative methods include

1. The Jacobi iterative method

2. The Gauss-Seidel iterative method

3. Relaxation methods: successive under-relaxation (SUR) and (SOR)

4. Iterative refinement for linear systems

5. The gradient and (preconditioned) conjugate gradient methods

These methods are best used for large sparse linear systems, since they are efficient in terms of both computation time and computer storage.

When utilizing iterative methods to solve a given $n \times n$ linear system $Ax = b$, an initial approximation $\mathbf{x}^{(0)}$ to the solution $\mathbf{x}$ is considered, which generates a sequence $\{\mathbf{x}^{(k)}\}_{k=0}^{\infty}$, such that, under certain conditions, it converges to $\mathbf{x}$. The fundamental idea behind these methods is the conversion of the system of interest into an equivalent system of the form $\mathbf{x} = T\mathbf{x} + \mathbf{c}$ for a fixed matrix T and a vector $\mathbf{c}$. After the initial vector $\mathbf{x}^{(0)}$ is selected, then the sequence of approximate solution vectors is generated by computing:

$$\mathbf{x}^{(k)} = T\mathbf{x}^{(k-1)} + \mathbf{c}, \text{ for each } k = 1, 2, 3, ... \tag{3.32}$$

Although these methods have been well documented, we will now briefly introduce the fundamental mechanics of a few of them.

The **Jacobi iterative method** starts by first considering $A = D - L - U$, where D is a diagonal matrix containing the diagonal entries of A, and $-L$ and $-U$ are the strictly lower-triangular and upper-triangular parts of A, respectively. For this substitution, we are able to rewrite the system of interest as

$$D\mathbf{x} = (L + U)\mathbf{x} + \mathbf{b}, \tag{3.33}$$

for which, if D^{-1} exists, the solution can then be computed with the iterative matrix equation

$$\mathbf{x}^k = D^{-1}(L + U)\mathbf{x}^{k-1} + D^{-1}\mathbf{b}, \quad k = 1, 2, \dots \tag{3.34}$$

The relation to (3.32) should be simple to appreciate.

A commensurate approach is embraced in the **Gauss–Seidel iterative method**, in which a similar iterative matrix equation allows for the solution to be attained, as follows:

$$\mathbf{x}^k = (D - L)^{-1}U\mathbf{x}^{k-1} + (D - L)^{-1}\mathbf{b}, \quad k = 1, 2, \dots \tag{3.35}$$

Iterative methods employ stopping criteria as a means of determining whether the solution at any given iteration is good enough, based upon a given tolerance. One common criterion is to stop when the **residual vector** at a given iteration k, defined and denoted as $\mathbf{r}^k \triangleq \mathbf{b} - A\mathbf{x}^k$, can be said to be less than a defined numerical tolerance ϵ; that is, the iteration will stop when $||\mathbf{r}^k|| < \epsilon$. Similar criteria could then be defined, which has to be implemented depending on the properties of the system to be solved.

3.5 Solution of Nonlinear Equations

When the PDE of interest is nonlinear, then the resulting algebraic system of equations $\mathbf{f}(\mathbf{x}) = \mathbf{0}$ is also nonlinear. The best-known method for solving linear equations is **Newton's method**. Let $f(x)$ be a scalar function, and let x^* be a root of f, i.e., $f(x^*) = 0$; after choosing an initial approximation, we compute

$$x_{i+1} = x_i - \frac{f(x_i)}{f'(x_i)}, \quad i = 0, 1, 2, \dots \tag{3.36}$$

Newton's method is a realization of the linearizing of $f(x)$, that is, it replaces $f(x)$ with a linear approximation.

Now, consider the provided initial estimate x_0, with which one can show that if $|x_0 - x^*|$ is small enough, and if f satisfies some (weak) conditions, then this process results in second-order convergence, i.e., $|x_{i+1} - x^*| = O(|x_{i+1} - x^*|^2)$.

The same concept can be used to solve a system of equations $\mathbf{f}(\mathbf{x}^*) = \mathbf{0}$, where $\mathbf{f}(\mathbf{x}) = (f_1(\mathbf{x}), ..., f_n(\mathbf{x}))^T$, $\mathbf{x} = (x_1, ..., x_n)^T$. Expanding $\mathbf{f}(\mathbf{x})$ around an approximation $\mathbf{x}_0$ of $\mathbf{x}^*$ gives the linear expressions

$$\mathbf{l}(\mathbf{x}) = \mathbf{f}(\mathbf{x}_0) + \mathbf{J}(\mathbf{x}_0)(\mathbf{x} - \mathbf{x}_0), \tag{3.37}$$

where

$$\mathbf{J}(\mathbf{x}_i) = \left(\frac{\partial \mathbf{f}_i}{\partial x_i} \right) = \begin{pmatrix} \frac{\partial \mathbf{f}_1}{\partial x_1} & \cdots & \frac{\partial \mathbf{f}_1}{\partial x_n} \\ \vdots & & \vdots \\ \frac{\partial \mathbf{f}_n}{\partial x_1} & \cdots & \frac{\partial \mathbf{f}_n}{\partial x_n} \end{pmatrix} \tag{3.38}$$

is the **Jacobian** of $\mathbf{f}(\mathbf{x})$. Therefore, given the initial approximation $\mathbf{x}_0$, the generalization of the previously presented Newtonian methods is

$$\mathbf{J}(\mathbf{x}_i)(\mathbf{x}_{i+1} - \mathbf{x}_i) = -\mathbf{f}(\mathbf{x}_i). \tag{3.39}$$

As in the prior scalar case, one can show that, if $||\mathbf{x}^* - \mathbf{x}_0||$ is small enough, and $\mathbf{f}$ satisfies the suitable conditions, the result is quadratic convergence, i.e.,

$$||x_{i+1} - \tilde{x}_i|| = O(||x_i - \tilde{x}_i||^2). \tag{3.40}$$

At each iteration, quadratic convergence can be achieved though actual computation; hence, Newton's method converges very quickly when $\mathbf{x}_0$ is sufficiently close to $\mathbf{x}^*$. This is a local property, so as the initial guess moves away from the actual solution, the method may converge slowly or not at all. There are many modifications of the method that have better global convergence properties, and are also more cost-effective to implement.

Quasi-Newton methods, including the **Broyden method**, exist to avoid the computational implications of having to compute the Jacobian matrix at each iteration; instead, it replaces it with an approximation matrix that is updated at each iteration.

Similarly, the **steepest descent technique** exists as a means of avoiding the disadvantage of a having a relatively accurate initial solution. This method converges even for poor initial estimates, thereby giving it the ability to determine a sufficiently accurate initial estimate for Newton's method.

3.6   Concluding Remarks

In this chapter, we introduced fundamental concepts in applied numerical analysis. We explained the importance of concepts such as the order of accuracy, and we explained important concepts in numerical linear algebra. Special attention was given to the problem system of equations, both linear and nonlinear.

3.7   Sample Problems

1. An important concept related to a given matrix $A \in \mathbb{C}^{m \times m}$, is that of its eigenvalues. Gershgorin's theorem reads, as follows:

   > **Theorem (Semyon Aranovich Gershgorin[1]).** Let $A \in \mathbb{C}^{m \times m}$. Every eigenvalue of A lies in at least one of the m circular disks in the complex plane with centers a_{ii} and radii
   >
   > $$\sum_{j \neq i} |a_{ij}|, \; j \in 1, ..., m. \tag{3.41}$$
   >
   > Moreover, if n of these disks form a connected domain that is disjointed from the other $m - n$ disks, then there are precisely n eigenvalues of A within this domain.

   (a) Prove Gershgorin's theorem.

   (b) Generate a plot of the eigenvalues and their respective bounding disk on the complex plane.

   (c) Consider an arbitrary matrix $A \in \mathbb{C}^{m \times m}$. For a given value of $m \geq 3$, what is the condition number of that matrix?

2. Consider: `A = hilb(n)` and `x = ones(n)`. Consider $\tilde{x} = x$ and $y = x$.

   (a) Compute $b = Ax$.

   (b) Solve $A\tilde{x} = b$ by means of LU factorization and compute the relative error

   $$\frac{||\tilde{x} - x||}{||x||}. \tag{3.42}$$

   for different values of n.

   (c) Generate a log plot of the attained norm for different values of n.

3. Consider `A = hilb(n)` and `x = ones(n)`. Consider $\tilde{x} = x$ and $y = x$. Compute $b = Ax$. Solve $A\tilde{x} = b$ by means of

   (a) Gauss–Seidel iterative method.

   (b) Jacobi iterative method.

   (c) SOR.

   For each method, generate a log plot of the attained 2-norm relative error for different values of n.

---

[1](August 24, 1901–May 30, 1933) from Pruzhany, Belarus.

4. Consider "A = gallery('lehmer',50);." Similarly, consider a variable defined in terms of a linearized distribution of nodes: "xx_true = linspace(1,50,50)';." Consider $\tilde{x} = x$ and $y = x$. Consider the following algorithms:

```
function [xx, iters, norms] =
  MyGradientMethod(AA, bb, epsilon, correct)

        % Initial guess:
        xx = rand(size(bb));
        % We will use the residual-based stopping
        % criteria:
        stop = false;
        iter = 1;
        while ~stop
                rr = bb - AA*xx;
                alpha = (rr'*rr)/(rr'*AA*rr);
                xx = xx + alpha*rr;
                if (norm(rr) <= epsilon)
                        stop = true;
                end
                iters(iter) = iter;
                norms(iter) =...
        norm(correct - xx)/norm(correct);
                iter = iter + 1;
        end
end

function [xx, iters, norms] =
  MyConjugateGradientMethod(AA, bb, epsilon, correct)

        % Initial guess:
        xx = zeros(size(bb));
        rr = bb;
        pp = rr;
        % We will use the residual-based stopping
        % criteria:
        stop = false;
        iter = 1;
        while ~stop
                alpha = (rr'*rr)/(pp'*AA*pp);
                xx = xx + alpha*pp;
                aux = rr;
                rr = rr - alpha*AA*pp;
                beta = (rr'*rr)/(aux'*aux);
                pp = rr + beta*pp;
```

```
              if  (norm(rr)/norm(bb) <= epsilon)
                     stop = true;
              end
              iters(iter) = iter;
              norms(iter) =...
        norm(correct - xx)/norm(correct);
              iter = iter + 1;
          end
end
```

Compute $b = Ax$. Solve $A\tilde{x} = b$ by means of

(a) Gradient method.

(b) Conjugate gradient method.

For each method, generate a log plot of the attained 2-norm relative error for different values of n. Can you compare both methods?

5. Consider Newton's method, which is based on the following iterative equation:

$$p_n = p_{n-1} - \frac{f(p_{n-1})}{f'(p_{n-1})}. \qquad (3.43)$$

(a) Solve $f(x) = \cos(x) - x$, $x \in [0, \pi/2]$ by means of a software package to get an accurate solution.

(b) Solve the proposed equation using Newton's method. Create a table to analyze the convergence of the solution as a function of the iteration number.

(c) Newton's method has an important drawback, and that is the necessity of computing the derivative of the function. Consider the following approximation instead:

$$f'(p_{n-1}) \approx \frac{f(p_{n-1}) - f(p_{n-2})}{p_{n-1} - p_{n-2}}, \qquad (3.44)$$

which yields the following iterative equation:

$$p_n = p_{n-1} - \frac{f(p_{n-1})(p_{n-1} - p_{n-2})}{f(p_{n-1}) - f(p_{n-2})}. \qquad (3.45)$$

Equation (3.45) is the core of the **Secant method**. Implement this solution in the proposed equation and compare by means of a table, its convergence versus Newton's method.

Chapter 4

Mimetic Differential Operators

Our study of partial differential equations for an unknown function u and the associated conservation laws (see Chapter 2) is the motivation for our subsequent development of a discretization numerical method for obtaining the discrete **operators** DIV, GRAD, and CURL. Note that this numerical method should have the property of approximating the continuous differential operators *div*, *grad*, and *curl* at the chosen grid points (or nodes) derived from the mesh of cells covering the physical domain.

The expanded form of Gauss' divergence theorem intimates the importance of considering two types of continuous functions **simultaneously**. The first type is a vector-valued function **v**, and the second type is some scalar-valued function f, so that *grad* f and *div* **v** have to be discretized using **convenient grids** for scalar and vector functions. The discretization then yields corresponding grid functions $\tilde{v}$ and $\tilde{f}$, which belong to vector spaces of different finite dimensions. We begin our study of lower-dimensional models and grids via the utilization of a 3-D general overview, as this has consistently proven itself to be the most effective approach.

A Cartesian mesh, obtained while discretizing the physical domain by means of Cartesian coordinates x, y, and z, is the simplest to visualize. The 3-D mesh elements, or cells, are convex polyhedra; in this case, rectangular parallelepipeds P, with one center M_3, six rectangular faces F, twelve edges, and eight 3-D vertices A, A^*, B, B^*, C, C^* and D, D^* (see Figure 4.1).

Two-dimensional problems arise when the considered model ignores the variation of relevant magnitudes along one of the three spatial directions, say the z-direction, and the distribution of u is an unknown function of x and y, exclusively. Keeping the original 3-D Cartesian mesh in the background, we see that the 2-D rectangular cells R are simply projections of the 3-D cells P upon the x-y plane, R having the four edges E projected from the four faces F of P, so that the 2-D edges are analogous to the 3-D faces. When it comes to applications, it helps to visualize such rectangles as very thin layers of fluid, such as the oily film that moves over soap bubbles.

In turn, the 1-D interval cells I are projections of the 2-D rectangular cells R, with two vertices V projected from two parallel edges of R. Therefore, the 1-D vertices are analogous to the 2-D edges and to the 3-D faces. It is often useful to visualize such intervals I as thin capillary vessels carrying vital fluids, such as sap, blood, and so forth; so that what we call "edges" or "vertices" in the 2-D and 1-D cases are really **faces**, in which lengths along the 1- or 2-dimensions have become negligible for the model at hand.

Centers project onto centers: $M_3 \longrightarrow M_2 \longrightarrow M_1$ (see Figure 4.1).

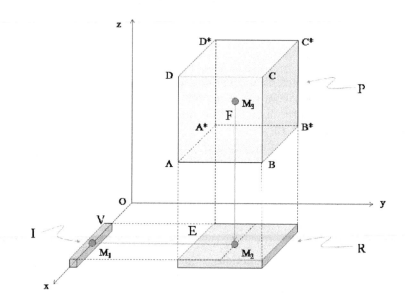

Figure 4.1: Global perspective of the 3-D cell concept and its lower-dimensional implications.

This mathematical analogy has a physical bearing, as can be seen from our previous modeling of the physical domain associated with a blood capillary vessel using a real interval $I = [0,1]$, so that if $V = (1,0,0)$, then this boundary point models a full 2-D face F, such as $ABCD$ (with unit area in the capillary model). By the same token, $(0,0,0)$ would model another 2-D face parallel to $ABCD$ but placed upon the plane $x = 0$, instead of the plane $x = 1$.

In Chapter 1, we emphasized that the Laplacian of U, or $(div\ grad\ U)$, plays an important role in the modeling of diverse physical conservation laws, so that we shall also be interested in the corresponding mimetic discrete analog.

Now, let us suppose that the scalar function acted upon by the operator $grad$ is f; i.e., $grad\ f$ is a vector function $\mathbf{v}$, and that we are also interested in the resulting scalar function $div\ \mathbf{v}$. As we shall see, the corresponding 3-D discrete operator DIV is best evaluated at cell centers, from the knowledge of values of $\mathbf{v}$ at the centers of the six rectangular faces F. If we disregard the z-variation, then we would need knowledge of $\mathbf{v}$ at the center of the four edges E; and, if we can further disregard the y-variation, we would need to know $\mathbf{v} = grad\ f$ at the vertices V of the projected interval $I = [0, 1]$ along the x-axis, that is, at $x = 0$ and $x = 1$, which are modeling very small faces, normal to the x-axis, and through which the field $\mathbf{v}$ can flow, **inwardly**, at $x = 0$, and **outwardly**, at $x = 1$ (see Figure 4.3).

Then the question arises as to where we need to know the values of the scalar function f, so that the 1-D discrete operator GRAD, corresponding to the continuous $grad$ and acting upon $\tilde{f}$, may act as a vector flux density at the vertices or boundary points of real intervals I in the 1-D case. For brevity's sake, in all dimensions, we shall write DIV $= \mathbf{D}$, GRAD $= \mathbf{G}$, and CURL $= \mathbf{C}$, where CURL denotes the discrete operator corresponding to $curl$.

Our answer is to know the values of the scalar f at the added grid points in a **staggered grid** superimposed on the original grid. The original grid results from a **mesh**, made up of cells covering the physical domain, and is constituted by **nodes**; i.e., by the intersection of three coordinate planes in the 3-D case, by the intersection of two coordinate lines in the 2-D case, and obtained as projections of the 2-D grid nodes in the 1-D case. These nodes can be **boundary nodes** or **inner nodes**. (Boundary nodes occur when these points belong to the boundary of the physical domain.)

In the case of a uniform, original 1-D grid, the nodes are $x_i = ih$, $0 \leq i \leq N$, where N is the number of 1-D cells and $h = 1/N$. A typical 1-D cell is the interval $[ih, (i+1)h]$. Boundary nodes are obtained for $i = 0$ and $i = N$.

The grid points **added** to the original grid through staggering are simply **the centers of the original 1-D cells**, that is to say, $x_{i+1/2} = (i + 1/2)h$, $0 \leq i \leq N - 1$. Graphic conventions for cell centers and original grid nodes will be discussed later in this chapter. Taken together, the inner nodes, cell centers, and boundary nodes comprise the complete set of grid points for a uniform staggered grid (see Figure 4.2).

In this way, the **simplest** discrete divergence and gradient are defined in a staggered grid by

$$(\mathbf{D}\tilde{\mathbf{v}})_{i+\frac{1}{2}} = \frac{v_{i+1} - v_i}{h}, \ 0 \leq i \leq N - 1 \tag{4.1}$$

$$(\mathbf{G}\tilde{f})_i = \frac{f_{i+\frac{1}{2}} - f_{i-\frac{1}{2}}}{h}, \ 1 \leq i \leq N - 1. \tag{4.2}$$

For boundary evaluations of $\mathbf{G}\tilde{f}$ at x_0 and x_N, we will also need to define the values f_0 and f_N, but in a way that preserves the global accuracy of the discrete method; this is accomplished by the Castillo–Grone method. The **simplest**, although not the most accurate, boundary evaluations for $\mathbf{G}\tilde{f}$, are given by

$$(\mathbf{G}\tilde{f})_0 = \frac{f_{\frac{1}{2}} - f_0}{\frac{h}{2}} \tag{4.3}$$

$$(\mathbf{G}\tilde{f})_N = \frac{f_N - f_{N-\frac{1}{2}}}{\frac{h}{2}}. \tag{4.4}$$

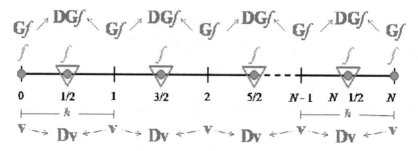

Figure 4.2: Staggered 1-D grid.

We will use the following notations to indicate what is going to be calculated at each location of the staggered grid (see Figure 4.2):

1. Vertical lines (|) for the calculation of the gradient $\mathbf{G}f$ and $\mathbf{v}$

2. ∇ for the calculation of the divergence $\mathbf{Dv}$

3. • for the calculation of the scalar function f

Note that this notation is convenient in the sense that the symbol ∇ is going to be placed at the points where the Laplacian $\nabla^2 f$ is calculated, in cases where $\mathbf{v} = \mathbf{G}f$.

Similarly, note that the discrete divergence acts on $\tilde{\mathbf{v}}$, with $\tilde{\mathbf{v}}$ evaluated **at the original nodes**, and $\mathbf{D}\tilde{\mathbf{v}}$ evaluated **at the nodes added by the staggering (cell centers)**, while the discrete gradient will act on $\tilde{f}$, with $\tilde{f}$ evaluated **at the nodes added by the staggering (cell centers)**, and $\mathbf{G}\tilde{f}$ evaluated at the original nodes, so that $\tilde{f} \in \mathbb{R}^{(N+2)}$, while $\tilde{\mathbf{v}} \in \mathbb{R}^{(N+1)}$, and $\mathbf{D}\tilde{\mathbf{v}} \in \mathbb{R}^{N}$, $\mathbf{G}\tilde{f} \in \mathbb{R}^{(N+1)}$ (see Figure 4.2).

Recall that $(grad \ \tilde{f})$ is usually a flux vector density, as we have seen with Fourier's law for heat flux and many other examples. This is the reason why it should be evaluated at the same grid points as $\tilde{\mathbf{v}}$, and that is the same reason why $\mathbf{DG}\tilde{f}$, the discrete analog of the operator *div grad*, acting on f, should be evaluated at the same grid points as $\mathbf{D}\tilde{\mathbf{v}}$ (see Figure 4.2).

Based upon these considerations, the means by which to generalize the procedure for 2- and 3-D staggered grids should be fairly apparent, and will be thoroughly examined later in this chapter.

For now, keep in mind that $\tilde{\mathbf{v}}$ and $\mathbf{G}\tilde{f}$, being flux vector densities, must both be evaluated at the same grid points, and those grid points must be analogous to the centers of the six faces F of P (in the 3-D case). These centers become the midpoints of the edges E of R in the 2-D case, and the vertices V of I in the 1-D case, as previously pointed out (see Figure 4.3).

In summary, you usually compute both *curl* $\mathbf{v}$ and $\mathbf{v} = grad \ f$ at the centers of the surface boundary pieces, which are

$$\left\{ \begin{array}{ll} \text{FACE CENTERS} & \text{(3-D SCENARIO)} \\ \text{EDGE CENTERS} & \text{(2-D SCENARIO)} \\ \text{VERTICES (NODES)} & \text{(1-D SCENARIO)} \end{array} \right. \qquad (4.5)$$

You then compute *div* **v** at the centers of the (solid) cells, which are

$$\left\{ \begin{array}{ll} \text{3-D CELL}(P)\text{CENTERS} & \text{(3-D SCENARIO)} \\ \mathbf{v} \cdot \mathbf{n} \text{ non-zero in 6 FACES } (F). \\ \text{2-D CELL}(R)\text{CENTERS} & \text{(2-D SCENARIO)} \\ \mathbf{v} = P\mathbf{i} + Q\mathbf{j} \Rightarrow \mathbf{v} \cdot \mathbf{n} \text{ non-zero in 4 FACES } (E). \\ \text{1-D CELL}(I)\text{CENTERS} & \text{(1-D SCENARIO)} \\ \mathbf{v} = v\mathbf{i} \Rightarrow \mathbf{v} \cdot \mathbf{n} \text{ non-zero in 2 FACES}(V). \end{array} \right. \qquad (4.6)$$

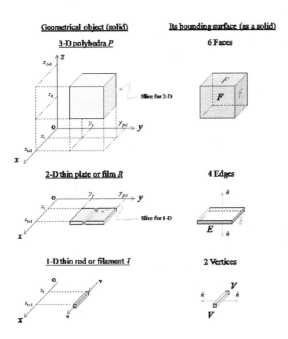

Figure 4.3: A dimension-wise view from top-to-bottom (sequential dimensional **collapse**).

Figure 4.3 shows, in the right column, that these faces may have different names, according to the dimensional scenario in which they are being considered; namely, faces, edges, or vertices.

4.1 Castillo–Grone Method for 1-D Uniform Staggered Grids

Besides approximating continuous differential operators, the constructed discrete analogs of these should also "mimic" specific properties and conditions, including the discrete fundamental theorem of calculus (FTC), and the discrete form of the integration by parts (IBP) formula. We also wish that **DG** approximates the continuous Laplace operator ∇^2.

Table 4.1: Discrete analogs of properties from continuous differential operators.

$\mathbf{D}\tilde{\mathbf{v}}_{const} = 0$	Free stream preservation
$\mathbf{G}\tilde{f}_{const} = 0$	Gradient of constant function is 0
$\mathbf{CG}\tilde{f} = 0$	Curl of the gradient is 0
$\mathbf{DC}\tilde{\mathbf{v}} = 0$	Div of curl is 0
$\mathbf{DG}\tilde{f} = \mathbf{L}f$	Laplacian is the divergence of the gradient

From Table 4.1, we adopt the following notations:

- With $\mathbf{y}^T$ being a $(1 \times N)$ matrix (row vector) and $\mathbf{Wx}$ a $(N \times 1)$ matrix (column vector), then we define: $< \mathbf{x}, \mathbf{y} >_{\mathbf{w}} \triangleq \mathbf{y}^T \mathbf{Wx}$, where $\mathbf{W}$ is a $N \times N$ weight square matrix.

- As usual, we define $< \mathbf{x}, \mathbf{y} > \triangleq x_1 y_1 + \cdots + x_N y_N \equiv \mathbf{y}^T \mathbf{x}$.

- $\mathbf{1}_N = (1, ..., 1)^T$ is an N-dimensional **column vector**, and $\mathbf{1}_N^T$ (the transpose of $\mathbf{1}_N$) is a **row vector**. When an N-tuple is displayed in the form $(x_1, ..., x_N)$, it is considered an N-dimensional **row vector**.

It is useful to recall that the continuous analog of the usual Euclidean inner product is

$$< f, g > = \int_0^1 f(x)g(x)dx, \tag{4.7}$$

and that the composite midpoint rule of integration is

$$\sum_{i=1}^{N} f_{i-\frac{1}{2}} h. \tag{4.8}$$

Table 4.2: Discrete integration analogs for the **simplest** discretized BCs for $\mathbf{G}\tilde{f}$.

---

1. Discrete (nodal grid) FTC: For $\tilde{\mathbf{v}} = (v_0, v_1, ..., v_N)^T \in \mathbb{R}^{(N+1)}$ and $\mathbf{D}\tilde{\mathbf{v}} \in \mathbb{R}^N$:

$$< \mathbf{D}\tilde{\mathbf{v}}, h\mathbf{1}_N >= v_N - v_0. \tag{4.9}$$

(composite midpoint rule of integration)

2. Discrete (cell-centered grid) FTC:

For $\tilde{f} = (f_0, f_{\frac{1}{2}}, f_{\frac{3}{2}} ..., f_{N-\frac{1}{2}}, f_N)^T \in \mathbb{R}^{(N+2)}$ and $\mathbf{G}\tilde{f} \in \mathbb{R}^{(N+1)}$:

$$< \mathbf{G}\tilde{f}, h\mathbf{1}_{N+1} >_P = h\mathbf{1}_{N+1}^T P\mathbf{G}\tilde{f} = f_N - f_0, \tag{4.10}$$

(trapezoid rule of integration)

with $P = DIAG\left(\frac{1}{2}, 1, 1, ..., 1, 1, \frac{1}{2}\right) \in \mathbb{R}^{(N+1)\times(N+1)}$.

3. Discrete (staggered grid) IBP:

$$h < \hat{\mathbf{D}}\tilde{\mathbf{v}}, \tilde{f} > + h < \mathbf{G}\tilde{f}, \tilde{\mathbf{v}} >_P = v_N f_N - v_0 f_0, \text{ or}$$
$$h < (\hat{\mathbf{D}} + (P\mathbf{G})^T)\tilde{\mathbf{v}}, \tilde{f} > = < \mathbf{B}\tilde{\mathbf{v}}, \tilde{f} >$$

with

$$\hat{\mathbf{D}} = \begin{bmatrix} 0 ... 0 \\ \mathbf{D} \\ 0 ... 0 \end{bmatrix} \text{ and } \mathbf{B} = \begin{bmatrix} -1 & 0 & ... & 0 & 0 \\ 0 & 0 & ... & 0 & 0 \\ & & \ddots & & \\ 0 & 0 & ... & 0 & 0 \\ 0 & 0 & ... & 0 & 1 \end{bmatrix}, \tag{4.11}$$

where both $\hat{\mathbf{D}}$ and $\mathbf{B}$ are elements of $\mathbb{R}^{(N+2)\times(N+1)}$. Notice that the extension of $\mathbf{D}$ to $\hat{\mathbf{D}}$ is necessary in order to have $\hat{\mathbf{D}}\tilde{\mathbf{v}} \in \mathbb{R}^{(N+2)}$, the same space to which $\tilde{f}$ belongs.

The Castillo–Grone method (CGM) (see [142], [143], [144], and [145]) allows for the construction of discrete differential operators which are k-th order accurate (**for any even** k), and **everywhere**, i.e., at the interior points and the domain boundary, **without using ghost points**. In order to achieve this goal of uniform accuracy, Castillo and Grone introduced **weighted inner products** with **weight matrices** $\hat{Q}$ and P, into their formulation, and operators $\overset{\circ}{\mathbf{D}}$ and $\check{\mathbf{G}}$ **adjusted** for the desired order k (see [142]):

$$h < \overset{\circ}{\mathbf{D}}\tilde{v}, \tilde{f} >_{\hat{Q}} +h < \check{\mathbf{G}}\tilde{f}, \tilde{\mathbf{v}} >_{P} = < \check{\mathbf{B}}\tilde{v}, \tilde{f} >, \qquad (4.12)$$

with $\hat{Q} \in \mathbb{R}^{(N+2)\times(N+2)}$ of the form

$$\hat{Q} = \begin{bmatrix} 0 \cdots 0 \\ Q \\ 0 \cdots 0 \end{bmatrix}, \ Q \in \mathbb{R}^{(N\times N)}. \qquad (4.13)$$

Similarly, $P \in \mathbb{R}^{(N+1)\times(N+1)}$ and

$$\check{\mathbf{B}} = h\{\hat{Q}\overset{\circ}{\mathbf{D}} + (P\check{\mathbf{G}})^{T}\}. \qquad (4.14)$$

The CGM constructs the uniformly k-th order accurate 1-D operator $\overset{\circ}{\mathbf{D}}$ in tandem with the weight matrix $\hat{Q}$, and the 1-D operator $\check{\mathbf{G}}$ with the weight matrix P. The resulting left-hand side of Equation (4.12) then forces a change to the original simple matrix $\mathbf{B}$ to some $(N + 2)$-by-$(N + 1)$ $\check{\mathbf{B}}$, as will be discussed further.

In the case of second-order operators, for which Q is the identity matrix, we obtain the same divergence mimetic operator $\mathbf{D}$ as the one produced by the support operators method (SOM) (see [65] and [145]). However, the gradient operator obtained by means of our method (CGM) is second-order throughout the entire domain (including the boundary); whereas, for the SOM, the accuracy of $\mathbf{G}$ on the boundary is one order less than the order achieved in the interior.

Fundamentally, the aim of CGM is to construct high-order approximations to **both** the divergence and gradient operators that satisfy a global conservation law, being high order both in the interior and on the boundary.

In this approach, (4.12) replaces the standard discretized version of the IBP; namely,

$$h < (Q\hat{\mathbf{D}} + (P\mathbf{G})^{T})\tilde{v}, \tilde{f} > = < \mathbf{B}\tilde{v}, \tilde{f} >, \qquad (4.15)$$

where the inner product formulas behave in reference to the usual Euclidean inner product, and $\mathbf{B}$ is the usual boundary operator (described by an $(N+2)$-by-$(N + 1)$ matrix), such that

$$\mathbf{B}\tilde{v} = (-v_0, 0, 0, ..., 0, v_N)^{T} \text{ and } (\mathbf{B}\tilde{v})^{T}\tilde{f} = < \mathbf{B}\tilde{v}, \tilde{f} > = v_N f_N - v_0 f_0, \quad (4.16)$$

This is the desired **exact net flux** through the boundary (in the x-direction), due to a flux vector density $\tilde{m}$, where $m_i = f_i v_i$, $0 \le i \le N$.

In order to pursue **higher-order** approximations, we must first define a natural **stencil matrix S**. The starting point for developing this concept is the notion of the stencil of a grid point X within a given mesh, which refers to a **geometrical arrangement** of a grid group with nonzero values attached to its grid elements; a group that relates to the point of interest X by using a numerical approximation scheme.

From this point on, we will now consider the **matrices** to be associated with our discrete operators **D**,**G**, and **C**, and, for brevity's sake, we will use the same symbol for the matrices associated with the operators. So, now $\mathbf{D} \in \mathbb{R}^{N \times (N+1)}$ and $\mathbf{D}v$ will stand for the matrix multiplication of $\mathbf{D}$ and the column matrix $v \in \mathbb{R}^{(N+1) \times 1}$. Analogously, $\mathbf{G}f_{cb}$ will stand for the matrix multiplication of $\mathbf{G} \in \mathbb{R}^{(N+1) \times (N+2)}$ and the column matrix $f_{cb} = \tilde{f} \in \mathbb{R}^{(N+2) \times 1}$.

In the **simplest** initial approach for **G**, from the formula

$$(\mathbf{G}f_{cb})_i = \frac{f_{i+\frac{1}{2}} - f_{i-\frac{1}{2}}}{h} = \frac{(-1)f_{i-\frac{1}{2}} + (1)f_{i+\frac{1}{2}}}{h}, \quad 1 \leq i \leq N-1, \quad (4.17)$$

we see that the stencil of the staggered grid point x_i resulting from the simplest operator $h\mathbf{G}$ is $(-1, 1)$.

For **D**, from the formula

$$(\mathbf{D}v)_{i+\frac{1}{2}} = \frac{v_{i+1} - v_i}{h} = \frac{(-1)v_i + (1)v_{i+1}}{h}, \quad 0 \leq i \leq N-1, \quad (4.18)$$

we see that the stencil of the staggered grid point $x_{i+1/2}$ resulting from operator $h\mathbf{D}$ is also $(-1, 1)$, but the matrix is now $N \times (N+1)$.

In this simple 1-D presentation, the stencils of the grid points which are the result of using discrete difference operators appear as interior **rows** of a global matrix, and this matrix exhibits a repetitive banded character, except possibly at the first and last rows, representing the stencils of the boundary nodes.

Before photocopiers were available, stencils were used to obtain duplicates of specific documents, by means of the visible materials being imprinted on the blank sheet in the duplicating machine. Analogously, the stencil of a matrix **M** describing a discrete differential operator arising from a given numerical method is the arrangement of matrix entries which differ from 0; and, while moving from one row to the next within the matrix, a **repetitive arrangement** becomes apparent.

In order to look after k-th order accurate divergences, gradients, and curls (k even), we first begin by utilizing the natural stencil matrix S. In the case of $k = 4$, the bandwidth of S equals 4, and the interior rows of the starting S have the form

$$\frac{1}{24}[0, ..., 0, 1, -27, 27, -1, 0, ..., 0]. \quad (4.19)$$

Such **canonical stencils** can be obtained using Lagrange polynomials (see [145]).

However, in order to achieve all of the desired mimetic properties for **D**, **G**, and **C**, we need the following additional properties of **D** (which will also be required for our initial stencils **S**):

1. $\mathbf{D}$ has zero row sums. Equivalently, $\mathbf{D}e = 0$, where $e = [1, 1, ..., 1]^T$.

2. $\mathbf{D}$ has column sums -1,0,...,0,1. Equivalently, $e^T\mathbf{D} = [-1, 0, 0, ..., 0, 1]$.

3. $\mathbf{D}$ is banded.

4. $\mathbf{D}$ has a "Toeplitz"-type structure on the interior rows **and is defined independently of** N, the number of grid points.

5. $\mathbf{D}$ is centro-skew-symmetric.

Conditions 1, 2, and 5 can be motivated as follows: Condition 1 is just $\mathbf{G}\tilde{f}_{const} = 0$. Condition 2 is equivalent to a discrete version of FTC with $\mathbf{v} \equiv 1$.

Conditions 1 and 2 must be **adjusted** in order for P and Q to drive the final form of the Castillo–Grone matrices $\mathring{D}$ and $\check{G}$.

The free stream preservation $div\ \mathbf{c} \equiv 0$ holding for constant $\mathbf{c}$, induces the following condition, to be satisfied by Q:

$$h < Q\mathbf{S_D}, e_f >\equiv< \mathbf{B}v, e_f >. \tag{4.20}$$

where $\mathbf{S}_D$ stands for the stencil yielding an initial $\mathbf{D}$, and $h\mathbf{S}_D = h\mathbf{D}$.

On the other hand, the FTC employing the weight matrix P now reads as follows:

$$h < e_v, P\mathbf{S_G}f >\equiv f_N - f_0. \tag{4.21}$$

This formula must be satisfied by $P \in \mathbb{R}^{(N+1)\times(N+1)}$. Here $e_v = 1_{N+1}^T = (1, ..., 1)$ is an $(N+1)$-dimensional row vector, while $e_f = 1_{N+2}$ is an $(N+2)$-dimensional column vector.

For a suitable matrix P, with structural conditions re-expressed in this manner (before, the condition for $\mathbf{S}$ was $e_v\mathbf{S} = [-1, 0, ..., 0, 1]$; now, the condition for $h\mathbf{S_G} = h\mathbf{G}$ is $P\mathbf{S_G} = [-1, 0, ..., 0, 1]$), the resulting linear systems for the P- and Q-entries are no longer incompatible, and, in fact, whole families of solutions can be found, as will be discussed later (see Appendix J).

Furthermore, once the final Castillo–Grone operators $\check{\mathbf{G}}$ and $\mathring{\mathbf{D}}$ are known, and the weight matrices P and Q are computed as described above, we can obtain the Castillo–Grone boundary operator $\check{\mathbf{B}}$, as follows:

$$\check{\mathbf{B}} = \hat{Q}h\mathring{\mathbf{D}} + (Ph\check{\mathbf{G}})^T. \tag{4.22}$$

The deduction of (4.22) is as follows: First, consider (4.12) so that, given the definition of weighted inner product, we will have

$$h < \hat{Q}\mathring{\mathbf{D}}v, f > +h < P\check{\mathbf{G}}f, v >=< \check{\mathbf{B}}v, f >$$
$$\Leftrightarrow\ < \hat{Q}\mathring{\mathbf{D}}v + (P\check{\mathbf{G}})^Tv, f >=< \frac{1}{h}\check{\mathbf{B}}v, f >$$
$$\Leftrightarrow\ \hat{Q}h\mathring{\mathbf{D}} + (Ph\check{\mathbf{G}})^T \equiv \check{\mathbf{B}},$$

where f is an $(N+2)$-dimensional column vector and v is a $(N+1)$-dimensional row vector.

Condition 5, i.e., skew-centro-symmetry, can be viewed as the algebraic equivalent of requesting that a real-valued function f of the real variable x, defined in the interval $[0,1]$, be skew-centro-symmetric; that is, f exhibits central symmetry with respect to the point $1/2$, a condition that forces the following slope property for the first ordinary derivative f': $f'(x) = f'(1-x)$ (see Figure 4.4).

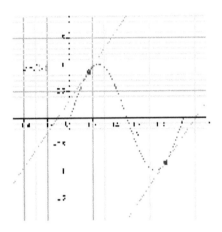

Figure 4.4: Centro-skew-symmetry.

Conditions 3 and 4 are useful for deriving efficient numerical algorithms from the mimetic schemes employing **D**, **G**, and **C**. By a Toeplitz-type structure we mean that the nonzero entries in row $(i+1)$ are just the nonzero entries of row i shifted one space to the right. In the case of a discrete gradient operator **G**, the idea is to extend a stencil to a mimetic gradient, ideally, $\mathbf{S} = h\mathbf{G}$. The Castillo–Grone strategy is to attempt to retain the stencils on the interior rows, constructing outer rows which satisfy the structural conditions 1, 2, and 5, while retaining the desired order of accuracy.

Conditions 1, 2, and 5 apply for both the mimetic divergence **D** and mimetic gradient **G**. For the simplest **G**, we take the interior rows for the matrix $h\mathbf{G}$, i.e., $[0\ 0\ \dots\ -1\ 1\ 0\ \dots\ 0]$, as the basic building block in a Toeplitz-type structure.

It is possible to construct **families** of uniformly fourth-order accurate gradients (see [142] and Appendix J), and, in particular, when we start with the interior rows given by (4.19), we obtain the following uniformly fourth-order accurate gradient:

$$
h\breve{\mathbf{G}} = \begin{pmatrix}
-\frac{1152}{407} & \frac{10063}{3256} & \frac{2483}{9768} & -\frac{3309}{3256} & \frac{2099}{3256} & -\frac{697}{4884} & 0 & \cdots \\
0 & -\frac{11}{12} & \frac{17}{24} & \frac{3}{8} & -\frac{5}{24} & \frac{1}{24} & 0 & \cdots \\
0 & \frac{1}{24} & -\frac{27}{24} & \frac{27}{24} & -\frac{1}{24} & 0 & 0 & \cdots \\
0 & 0 & \frac{1}{24} & -\frac{27}{24} & \frac{27}{24} & -\frac{1}{24} & 0 & \cdots
\end{pmatrix}. \tag{4.23}
$$

Therefore, the desired $(N+1)$-by-$(N+2)$ matrix $\breve{\mathbf{G}}$ uses local special formulas near the boundaries (affecting the two upper rows and the two lower rows), as well as the standard fourth-order approximation away from the boundary:

$$
(\breve{\mathbf{G}}f)_i = \frac{1}{h}\left(\frac{1}{24}f_{i-\frac{3}{2}} - \frac{27}{24}f_{i-\frac{1}{2}} + \frac{27}{24}f_{i+\frac{1}{2}} - \frac{1}{24}f_{i+\frac{3}{2}}\right). \tag{4.24}
$$

For k-th order accurate $\breve{\mathbf{G}}$, the weighting matrix $\breve{P}$ has the form

$$
\breve{P} = \tilde{P} \oplus I_{N-2k} \oplus J_k\tilde{P}J_k, \tag{4.25}
$$

where $\tilde{P}$ is a $k \times k$ **positive** definite matrix, and J_k denotes the $k \times k$ matrix containing ones on the main cross diagonal and zero elsewhere.

Analogously, for n-th order accurate $\mathring{\mathbf{D}}$, the weighting matrix has the form

$$
Q = \Lambda \oplus I_{N-2n} \oplus \Lambda', \tag{4.26}
$$

with $\Lambda' = P_n\Lambda P_n$, and Λ is a $n \times n$ positive definite matrix and $\mathbf{P}_n$ denotes the permutation matrix $n \times n$:

$$
\mathbf{P}_n = \begin{bmatrix}
0\,0 & \cdots & & 0 & 0\,1 \\
0\,0 & \cdots & & 0 & 1\,0 \\
0\,0 & \cdots & & 1 & 0\,0 \\
\vdots & & 0 & 1 & 0 & & \vdots \\
0\,0 & 1 & 0 & \cdots & 0\,0 \\
0\,1 & 0 & \cdots & & 0\,0 \\
1\,0 & 0 & \cdots & & 0\,0
\end{bmatrix}. \tag{4.27}
$$

Consider two 7×7 matrices

$$
\mathbf{M} = \begin{bmatrix}
a\,b\,0\,0\,0\,0\,0 \\
0\,c\,d\,0\,0\,0\,0 \\
0\,0\,0\,0\,0\,0\,0 \\
0\,0\,0\,0\,0\,0\,0 \\
0\,0\,0\,0\,0\,0\,0 \\
0\,0\,0\,0\,e\,f\,0 \\
0\,0\,0\,0\,0\,g\,h
\end{bmatrix} \text{ and } \mathbf{P}_7\mathbf{M}\mathbf{P}_7 = \begin{bmatrix}
h\,g\,0\,0\,0\,0\,0 \\
0\,f\,e\,0\,0\,0\,0 \\
0\,0\,0\,0\,0\,0\,0 \\
0\,0\,0\,0\,0\,0\,0 \\
0\,0\,0\,0\,0\,0\,0 \\
0\,0\,0\,0\,d\,c\,0 \\
0\,0\,0\,0\,0\,b\,a
\end{bmatrix}. \tag{4.28}
$$

Our motivation for introducing this permutation matrix arises from the consideration of a special, vector-valued function $\mathbf{v}(x) = v(x)\mathbf{i}$, with $0 \le x \le 1$, with a **skew-centrosymmetric graph**, so that v has the special property

$$
v(1-x) = -v(x), \tag{4.29}
$$

and $div \ \mathbf{v}(1 - x) = div \ \mathbf{v}(x)$, where $div \ \mathbf{v}(1 - x)$ means $div \ \mathbf{v}$ evaluated at $(1 - x)$. The condition $v(1 - x) = -v(x)$, defining skew-centrosymmetry for graphs, can be naturally adapted in the context of matrices.

If the **stencil** of matrix $\mathbf{M}$ is regarded as the analog of the graph for our special function v, then we will require for such a special $\mathbf{M}$, the obvious conditions: $e = -d$, $f = -c$, $g = -b$ and $h = -a$; then it holds that $P_7\mathbf{M}P_7 = -\mathbf{M}$.

In general matrix terms, this imposes the following analog symmetry condition on matrix $\mathbf{D}$: If $\mathbf{D}$ is an $N \times (N + 1)$ matrix, then the mimetic version of $div \ \mathbf{v}(1 - x) = div \ \mathbf{v}(x)$ is $P_N\mathbf{D}P_{N+1} = -\mathbf{D}$, and, according to our previous slope considerations for the graph of such a special $v(x)$, any $N \times (N + 1)$ matrix $\mathbf{M}$ satisfying $P_N\mathbf{M}P_{N+1} = -\mathbf{M}$, according to Andrew (see [146]), will be referred to as **centro-skew-symmetric**. The stencil matrices for the **simplest** operator $\mathbf{D}$ considered at the beginning are obviously $N \times (N + 1)$ centro-skew-symmetric matrices, and we desire to preserve this as a mimetic condition for the CGM matrix operators.

For the uniformly fourth-order accurate $\check{\mathbf{G}}$ presented above, $\tilde{P}$ is a 4×4 diagonal matrix given by

$$
\tilde{P} = \begin{bmatrix} \frac{407}{1152} & 0 & \cdots & 0 \\ 0 & \frac{473}{384} & 0 & \vdots \\ \vdots & 0 & \frac{343}{384} & 0 \\ 0 & \cdots & 0 & \frac{1177}{1152} \end{bmatrix}, \tag{4.30}
$$

for which

$$
\check{P} = \begin{bmatrix}
\frac{407}{1152} & 0 & 0 & 0 & 0 & 0 & \cdots & & \cdots & 0 \\
0 & \frac{473}{384} & 0 & 0 & 0 & 0 & \cdots & & \cdots & 0 \\
0 & 0 & \frac{343}{384} & 0 & 0 & 0 & \cdots & & \cdots & 0 \\
0 & 0 & 0 & \frac{1177}{1152} & 0 & 0 & \cdots & & \cdots & 0 \\
0 & 0 & 0 & 0 & 1 & 0 & \cdots & & \cdots & 0 \\
0 & \cdots & & & & \ddots & & & \cdots & 0 \\
0 & \cdots & & & \cdots & 0 & 1 & 0 & 0 & 0 & 0 \\
0 & \cdots & & & \cdots & 0 & 0 & \frac{1177}{1152} & 0 & 0 & 0 \\
0 & \cdots & & & \cdots & 0 & 0 & 0 & \frac{343}{384} & 0 & 0 \\
0 & \cdots & & & \cdots & 0 & 0 & 0 & 0 & \frac{473}{384} & 0 \\
0 & \cdots & & & \cdots & 0 & 0 & 0 & 0 & 0 & \frac{407}{1152}
\end{bmatrix}. \tag{4.31}
$$

Also,

$$
Q = \begin{bmatrix}
\frac{1705}{6156} & 0 & 0 & 0 & 0 & 0 & \cdots & & \cdots & 0 \\
0 & \frac{5995}{4104} & 0 & 0 & 0 & 0 & \cdots & & \cdots & 0 \\
0 & 0 & \frac{1363}{2052} & 0 & 0 & 0 & \cdots & & \cdots & 0 \\
0 & 0 & 0 & \frac{13519}{12312} & 0 & 0 & \cdots & & \cdots & 0 \\
0 & 0 & 0 & 0 & 1 & 0 & \cdots & & \cdots & 0 \\
0 & \cdots & & & & & \ddots & & \cdots & 0 \\
0 & \cdots & & & & \cdots & 0 & 1 & 0 & 0 & 0 & 0 \\
0 & \cdots & & & & \cdots & 0 & 0 & \frac{13519}{12312} & 0 & 0 & 0 \\
0 & \cdots & & & & \cdots & 0 & 0 & 0 & \frac{1363}{2052} & 0 & 0 \\
0 & \cdots & & & & \cdots & 0 & 0 & 0 & 0 & \frac{5995}{4104} & 0 \\
0 & \cdots & & & & \cdots & 0 & 0 & 0 & 0 & 0 & \frac{1705}{6156}
\end{bmatrix}. \qquad (4.32)
$$

In the CGM, the $(N+1)$-by-$(N+1)$ weight matrix $\check{P}$ was originally introduced to ensure the solvability of the linear system imposing the desired conditions 1 through 5, thus motivating the following form:

$$
\check{P} = \tilde{P} \oplus I_{N-2k} \oplus J_k \tilde{P} J_k, \qquad (4.33)
$$

however; in order to pursue the final form for accurate discrete operators, the desired row and column sum conditions must be recast with respect to the weighted inner product $< .,. >_{\check{P}}$, as previously explained. For simplicity's sake, in the case of the weight matrix $\check{P}$ adjoined to $\check{\mathbf{G}}$, we write $\check{P} = P$.

With the appropriate CG matrix P, the resulting algebraic linear system is no longer incompatible; and, in fact, as was previously pointed out, entire families of solutions can be produced for each even order k of desired **uniform accuracy**, with $P(h\check{\mathbf{G}})$ satisfying the row and column sum requirements, besides being centro-skew-symmetric (see Appendix J).

An alternative approach to that presented in [142] for the construction of Q, $\hat{\mathbf{D}}$, P and, $\check{\mathbf{G}}$ is presented in [147]. A **simpler** way of constructing compatible algebraic linear systems satisfying a recast row and column sum conditions was given in [144]. In the case of a fourth-order accurate 1-D divergence, this simpler method will be outlined in Appendix J.

It is important to note that Q, and, therefore, the augmented $\hat{Q}$, both depend on k **and are independent of** N, but the number of grid points must be sufficiently large for each desired k, and, as is shown in [142], $N \geq 3k - 1$ must also hold true.

4.2   Higher-Dimensional CGM

The Castillo–Grone (CG) 2-D divergence and gradient mimetic operators defined on a 2-D uniform staggered grid were first presented in [21], and the

2-D discrete curl operator was introduced in [144], in a work that also improves the matrix representation of the 2-D divergence and gradient mimetic operators by means of a numbering system; thereby allowing a more compact version of these operators, resulting from Kronecker products of block matrices.

Now, recall that the lexicographical order on $X \times Y$, the Cartesian product of sets X and Y, is defined as

$$(x_a, y_b) < (x'_a, y'_b) \iff x_a < x'_a \text{ or } (x_a = x'_a \text{ and } y_b < y'_b), \qquad (4.34)$$

for any (x_a, y_b) and $(x'_a, y'_b) \in X \times Y$.

On the other hand, $\mathbf{A} \otimes \mathbf{B}$, the Kronecker product of an $m \times n$ matrix $\mathbf{A} = a_{ij}$ with a $p \times q$ matrix $\mathbf{B}$, is the $mp \times nq$ matrix given by

$$\mathbf{A} \otimes \mathbf{B} = \begin{bmatrix} a_{11}\mathbf{B} & \cdots & a_{1n}\mathbf{B} \\ \vdots & \ddots & \vdots \\ a_{m1}\mathbf{B} & \cdots & a_{mn}\mathbf{B} \end{bmatrix}. \qquad (4.35)$$

Now, letting $\mathbf{D}_N \in \mathbb{R}^{N \times (N+1)}$ be a discrete first derivative operator that maps from the $(N+1)$ nodes of a 1-D uniform staggered grid to the N cell centers; and $\hat{\mathbf{D}}_N \in \mathbb{R}^{(N+1) \times N}$ be the map that approximates the first derivative at the $(N-1)$ inner nodes from values at the N cell centers and assigns the value zero at the two boundary points, then the 2-D CG curl operator developed in [144] is given by

$$h\mathbf{C} = \begin{bmatrix} \mathbf{0} & \mathbf{0} & \hat{D}_N \otimes I_M \\ \mathbf{0} & \mathbf{0} & -I_N \otimes \hat{D}_M \\ -D_N \otimes I_M & I_N \otimes D_M & \mathbf{0} \end{bmatrix}. \qquad (4.36)$$

Here, $\mathbf{D}$ and its corresponding augmented $\hat{\mathbf{D}}$ are arbitrary, even-order accurate first derivative operators constructed using the CGM.

Unlike the abovementioned approach to the 2-D curl by means of line integrals or circulations around a closed loop, here we use a simpler, unified higher-dimensional approach (for both 2- and 3-D) for curls and gradients as limiting fluxes-per-unit-volume in the limit as the volume approaches zero, thus avoiding the consideration of diverse vector spaces by means of adequate staggerings.

The generalized coordinate-free formulation of the DEL operator as a limiting shrinking FLUX per unit volume is

$$\nabla(\ )\phi = \lim_{\text{Diam } \Omega \to 0} \frac{\int_\sigma \mathbf{n}(\ )\phi \, d\sigma}{\text{Vol } \Omega}. \qquad (4.37)$$

In the above formulation, Ω can be taken as a 3-D cell P with center M_3, and σ is its bounding surface; in particular, it is the six faces F (see Figure 4.1).

Vol Ω is the measure of Ω, i.e., its volume in the 3-D case, and Diam Ω is its diameter.

In this universal DEL formula, one must replace ( ) by $\cdot$, $\times$, or nothing, depending upon the field ϕ being operated upon (vector- or scalar-valued).

In particular:

$$\nabla \times \mathbf{v}(M_3) = \lim_{\text{Diam } \Omega \to 0} \frac{\int_\sigma \mathbf{n} \times \mathbf{v} \, d\sigma}{\text{Vol } \Omega}. \qquad (4.38)$$

Since the **triple scalar product** $[\mathbf{e} \, \mathbf{n} \, \mathbf{v}] = [\mathbf{n} \, \mathbf{v} \, \mathbf{e}] = <\mathbf{n}, \mathbf{v} \times \mathbf{e}>$, where $\mathbf{e}$ is any unit vector along one of the three Cartesian axes, then we immediately see that any of the three scalar components of $\nabla \times \mathbf{v}(M_3)$ can be discretized at the stencil center M_3, inspired by

$$< (\nabla \times \mathbf{v})(M_3), \mathbf{e} >= \lim_{\text{Diam } \Omega \to 0} \frac{\int_\sigma (\mathbf{v} \times \mathbf{e}) \cdot (\mathbf{n} \, d\sigma)}{\text{Vol } \Omega}. \qquad (4.39)$$

Therefore, it is sufficient to have 2-D mimetic divergence discrete operators with the desired even order of accuracy, coupled with adequate 2-D or 3-D staggerings, in order to obtain the single scalar component of CURL $\mathbf{v}$ in 2-D, or the three scalar components of CURL $\mathbf{v}$ in 3-D, as we shall presently see.

4.3   2-D Staggerings

In this section, we present important considerations for extending the operators to two-dimensional problems.

4.3.1   2-D Gradient

For a 2-D scalar field f, we have the continuous operator *grad*, such that

$$grad \, f(x, y) = \frac{\partial f}{\partial x}\mathbf{i} + \frac{\partial f}{\partial y}\mathbf{j}, \qquad (4.40)$$

and the flux-motivated formula is

$$grad \, f(x, y) = \lim_{\text{Diam } R \to 0} \frac{\int_R \mathbf{n} f \, dxdy}{\text{Area } R}, \qquad (4.41)$$

with our previous notations already used for *div* $\mathbf{v}(x, y)$.

Therefore, for the purpose of constructing a 2-D stencil, we choose to evaluate GRAD $f(x_i, y_{j+1/2})$, where the latter is defined by (see Figure 4.5)

$$\left[\frac{f(i + \frac{1}{2}, j + \frac{1}{2}) - f(i - \frac{1}{2}, j + \frac{1}{2})}{(x_{i+\frac{1}{2}} - x_{i-\frac{1}{2}})}\mathbf{i} + \frac{f(i, j + 1) - f(i, j)}{(y_{j+1} - y_j)}\mathbf{j} \right]. \qquad (4.42)$$

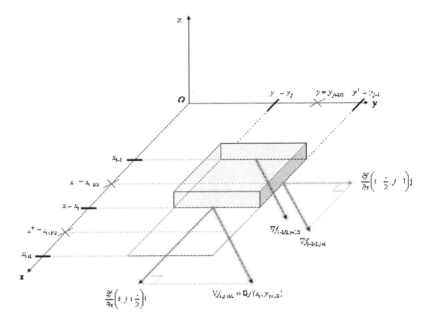

Figure 4.5: 2-D stencil for both partial and full 2-D staggerings for GRAD f.

Using the preceding stencil, we are able to calculate $\mathbf{G}f$ at the centers of the edges of the original grid, by means of combined x and y-staggerings, as exhibited in Figure 4.6.

For the simplest numerical approximation of $\mathbf{G}f$ we have

1. With f at □-denoted points (i,j), $(i+1,j)$, and at ∇-denoted points $(1+1/2, j-1/2)$ and $(i+1/2, j+1/2)$, we can obtain $\mathbf{G}f$ at the diamond-denoted points $(i+1/2, j)$.

2. With f at □-denoted points $(i, j-1)$, $(i+1, j-1)$, and at ∇-denoted points $(i+1/2, j-3/2)$ and $(i+1/2, j-1/2)$, we can obtain $\mathbf{G}f$ at the diamond-denoted point $(i+1/2, j-1)$.

3. With f at □-denoted points $(i-1, j)$, (i, j), and at ∇-denoted points $(i-1/2, j-1/2)$ and $(i-1/2, j+1/2)$, we can obtain $\mathbf{G}f$ at the diamond-denoted point $(i-1/2, j)$.

4. With f at □-denoted points $(i-1, j-1)$, $(i, j-1)$, and at ∇-denoted points $(i-1/2, j-3/2)$ and $(i-1/2, j-1/2)$, we can obtain $\mathbf{G}f$ at the diamond-denoted point $(i-1/2, j-1)$.

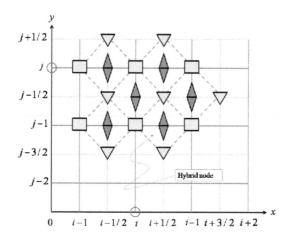

Figure 4.6: Partial 2-D staggering for 2-D GRAD f.

5. With f at $\square$-denoted points $(i, j-1)$, (i,j), and at ∇-denoted points $(i-1/2, j-1/2)$ and $(i+1/2, j-1/2)$, we can obtain $\mathbf{G}f$ at the diamond-denoted point $(i, j-1/2)$.

6. With f at $\square$-denoted points $(i+1, j-1)$, $(i+1,j)$, and at ∇-denoted points $(i+1/2, j-1/2)$ and $(i+3/2, j-1/2)$, we can obtain $\mathbf{G}f$ at the diamond-denoted point $(i+1, j-1/2)$.

For clarity's sake, let us define the **original grid** as the one containing only nodes with integer-indexed coordinates, which was originally defined before the introduction of staggered nodes.

Notice from the previous listing and from Figure 4.6 that the diamond-denoted nodes can be thought of as **hybrid nodes**, which are always edge centers of the original grid. Their hybrid character relies on the fact that if one of its coordinates belongs to the original grid, then the other coordinate will belong to the set of added staggered nodes.

From the previous deduction, and, as can be seen in [144], we can define a Castillo–Grone 2-D gradient operator, as follows:

$$\breve{\mathbf{G}} \triangleq \begin{bmatrix} \mathbf{G}_x \\ \mathbf{G}_y \end{bmatrix}, \tag{4.43}$$

where

$$\mathbf{G}_x \triangleq I_N^T \otimes \breve{\mathbf{G}}_M, \tag{4.44}$$

$$\mathbf{G}_y \triangleq \breve{\mathbf{G}}_M \otimes I_M^T. \tag{4.45}$$

4.3.2 2-D Divergence

For a 2-D vector field of the form $\mathbf{v} = P(x,y)\mathbf{i} + Q(x,y)\mathbf{j}$, the continuous operator is such that

$$div\ \mathbf{v} \triangleq \frac{\partial P}{\partial x} + \frac{\partial Q}{\partial y}, \tag{4.46}$$

and the flux-motivated formula is

$$div\ \mathbf{v}(x,y) = \lim_{\text{Diam } R \to 0} \frac{\int_R \mathbf{n} \cdot \mathbf{v}\ dxdy}{\text{Area } R}, \tag{4.47}$$

with the notations used above.

Therefore,

$$div\ \mathbf{v}(x,y) \approx \text{DIV}\ v\,(i+1/2, j+1/2), \tag{4.48}$$

where the latter is defined by

$$\left[\frac{P(i+1, j+\frac{1}{2}) - P(i, j+\frac{1}{2})}{(x_{i+1} - x_i)} + \frac{Q(i+\frac{1}{2}, j+1) - Q(i+\frac{1}{2}, j)}{(y_{j+1} - y_j)} \right]. \tag{4.49}$$

The numerical DIV v is obtained at the 2-D cell center $(x_{i+1/2}, y_{j+1/2})$, from the flux of $\mathbf{v} = P\mathbf{i} + Q\mathbf{j}$ through the edges of R, using the values of $\mathbf{v}$ at the corresponding edge centers (see Figure 4.7).

Similarly, as seen in [144] and related to (4.43), we can define a Castillo–Grone 2-D divergence operator, as follows:

$$\check{\mathbf{D}} \triangleq \begin{bmatrix} \mathbf{D}_x & \mathbf{D}_y \end{bmatrix}, \tag{4.50}$$

where:

$$\mathbf{D}_x \triangleq I_N \otimes \check{\mathbf{D}}_M, \tag{4.51}$$

$$\mathbf{D}_y \triangleq \check{\mathbf{D}}_N \otimes I_M. \tag{4.52}$$

4.3.3 2-D curl

In order to prepare the way for presenting the treatment of the 3-D CURL, we shall first exhibit our method for the simpler 2-D CURL.

Let $\mathbf{v}(x,y) = P(x,y)\mathbf{i} + Q(x,y)\mathbf{j}$, and define the auxiliary vector field $\mathbf{v}^*_{xy} = \mathbf{v} \times \mathbf{k} = Q\mathbf{i} - P\mathbf{j} = P^*\mathbf{i} + Q^*\mathbf{j}$, i.e., $P^* = Q$ and $Q^* = -P$. Since

$$curl\ \mathbf{v} = \mathbf{k}\left(\frac{\partial Q}{\partial x} - \frac{\partial P}{\partial y} \right), \tag{4.53}$$

then

$$\mathbf{k} \cdot curl\ \mathbf{v} = \frac{\partial P^*}{\partial x} + \frac{\partial Q^*}{\partial y} = div\ \mathbf{v}^*_{\mathbf{xy}}, \tag{4.54}$$

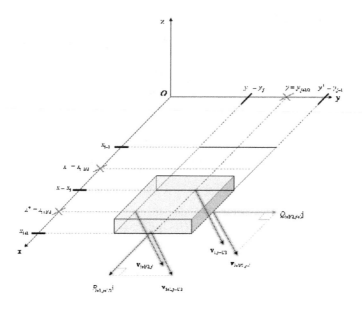

Figure 4.7: 2-D staggering for *div* **v**.

so that we only need to compute a 2-D discrete divergence at a point, in order to obtain a numerical approximation for the single scalar component of *curl* **v** there.

Since staggered grids will be considered, it is important to introduce a notational change from Figure 4.3 to Figure 4.8. The pairs $(x_{i-1/2}, x_{i+1/2})$, (y_j, y_{j+1}) will be renamed (x^-, x^+) and (y^-, y^+), respectively, and

$$y = y_{j+\frac{1}{2}} = \frac{y^- + y^+}{2} = y^0, \tag{4.55}$$

$$x = x_i = \frac{x^- + x^+}{2} = x^0. \tag{4.56}$$

The family of planes $x = x_{i-1/2} = x^-$ and $x = x_{i+1/2} = x^+$, together with the family of planes $y = y_{j-1/2}$ and $y = y_{j+1/2} = y^0$, which are added via staggering, form **a new** 3-D cell P^* with center M_3^*, and rectangular slices R^*, which are parallel to the plane $z = 0$.

This staggering, which will be coupled with the CG 2-D DIV operator, is motivated by the flux formula for the scalar components of *curl* **v**:

$$< (\nabla \times \mathbf{v})(M_3), \mathbf{k} > = \lim_{\text{Diam } \Omega \to 0} \frac{\int_\sigma \mathbf{n} \cdot (\mathbf{v} \times \mathbf{k}) \, d\sigma}{\text{Vol } \Omega}. \tag{4.57}$$

Note that this formula still holds true for a 3-D curl for $\mathbf{v} = p\mathbf{i} + q\mathbf{j} + r\mathbf{k}$, since $\mathbf{v} \times \mathbf{k} = q\mathbf{i} - p\mathbf{j}$.

Since $\partial \mathbf{v}/\partial z \equiv 0$ in the 2-D case, we can consider that the field $\mathbf{v}$ is being observed at any horizontal plane $z = z_k$; particularly, $z = 0$, as in Figure 4.1. We can also assume that the rectangular slice R^* has a small but finite width dz, and can be assimilated to a thin solid polyhedron (dP^*), with a volume equal to $dz(\text{Area } R^*)$.

Letting $\Omega = P^*$, so that the center of the 3-D cell P^*, formerly denoted as M_3^*, is projected as M_2^*; M_2^* is now renamed $(x, y, 0)$, or is simply (x, y). As far as M_2^* goes, it then becomes the center of R, where R denotes the projection of R^* onto the plane $z = 0$. For simplicity's sake, let us denote the four vertices of R as

$$A = (x^+, y^-), \ A^* = (x^-, y^-), \ B = (x^+, y^+) \ \text{and} \ B^* = (x^-, y^+). \quad (4.58)$$

Also, set $\Delta x = x^+ - x^- = x_{i+\frac{1}{2}} - x_{i-\frac{1}{2}}$, and $\Delta y = y^+ - y^- = y_{j+1} - y_j$.

Inspired by the formula

$$div \ \mathbf{v}^*(x, y) = \lim_{\text{Diam } R \to 0} \frac{\int_R \mathbf{n} \cdot \mathbf{v}^* \, dxdy}{\text{Area } R}, \quad (4.59)$$

which was obtained from the general 3-D formula by cancellation of the factor dz, and

$$\text{Vol}(dP^*) = dz \ (\text{Area } R), \quad (4.60)$$

where Area $R = \Delta x \Delta y$, since $curl \ \mathbf{v}(x, y) = \mathbf{k} \ div \ \mathbf{v}_{xy}^*$, we obtain the following expressions as the simplest numerical evaluations for $curl \ \mathbf{v}(x, y) = curl \ \mathbf{v}(x_i, y_{j+\frac{1}{2}})$, to this low order of accuracy:

$$curl \ v(x, y) \approx \text{CURL } v\left(i, j + \frac{1}{2}\right) = \mathbf{k} \ \text{DIV } \mathbf{v}^*\left(i, j + \frac{1}{2}\right), \quad (4.61)$$

where

$$\text{CURL } v\left(i, j + \frac{1}{2}\right) \cdot \mathbf{k} = \left[\frac{Q(x^+, y) - Q(x^-, y)}{\Delta x} - \frac{P(x, y^+) - P(x, y^-)}{\Delta y}\right], \quad (4.62)$$

and for DIV v^*, evaluated at the same point, we have

$$\left[\frac{P^*(i + \frac{1}{2}, j + \frac{1}{2}) - P^*(i - \frac{1}{2}, j + \frac{1}{2})}{(x_{i+\frac{1}{2}} - x_{i-\frac{1}{2}})} + \frac{Q^*(i, j + 1) - Q^*(i, j)}{(y_{j+1} - y_j)}\right]. \quad (4.63)$$

Figure 4.9 depicts this.

Although this is a low-order approximation, it is useful to illustrate **the type of 2-D staggering** needed in order to compute $curl \ \mathbf{v}$ at any staggered grid point (x, y) from values of $\mathbf{v} = P\mathbf{i} + Q\mathbf{j}$ or $\mathbf{v}_{xy}^* = \mathbf{v} \times \mathbf{k} = Q\mathbf{i} - P\mathbf{j} =$

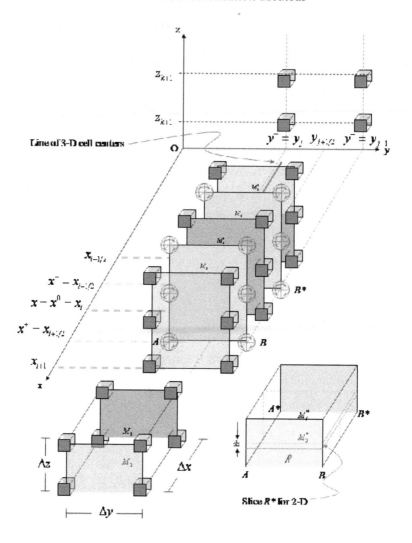

Figure 4.8: 3-D staggering for $\mathbf{k} \cdot curl \ \mathbf{v}$.

$P^*\mathbf{i}+Q^*\mathbf{j}$ at neighboring points (x^+, y), (x^-, y) for component Q, and at points (x, y^+), (x, y^-) for component P of $\mathbf{v}$; more importantly, it immediately allows us to visualize the staggerings needed when adopting the more sophisticated 2-D stencils of the CGM, which are uniformly even-order accurate, as shown earlier.

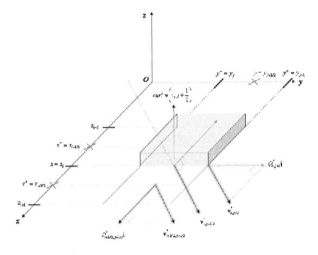

Figure 4.9: 2-D staggering for $\mathbf{k} \cdot curl\ \mathbf{v} = div\ \mathbf{v}^*$.

4.4 3-D Staggerings

Analogously, as for the two-dimensional case, in this section we present important considerations for extending the operators for use in solving three-dimensional problems.

4.4.1 3-D Gradient

Since

$$grad\ f(x, y, z) = \left(\frac{\partial f}{\partial x}\mathbf{i} + \frac{\partial f}{\partial y}\mathbf{j} + \frac{\partial f}{\partial z}\mathbf{k} \right) (x, y, z), \qquad (4.64)$$

and each scalar component is a first-order derivative, we need only 1-D divergences for numerically approximating each of them, and the required 3-D staggering should be obvious now, being so much simpler than those required for obtaining a 3-D CURL.

In this case, the three auxiliary vectors to be derived from the scalar function f in order to evaluate those components as 1-D divergences are

$$\mathbf{v}_x^* = f\mathbf{i},\ \mathbf{v}_y^* = f\mathbf{j}\ \text{and}\ \mathbf{v}_z^* = f\mathbf{z}. \qquad (4.65)$$

In this way,

$$div\ \mathbf{v}_x^* = \frac{\partial f}{\partial x} = \mathbf{i} \cdot grad\ f, \qquad (4.66)$$

and so forth.

4.4.2 3-D Divergence

Motivated by the definition presented in (4.49), the definition of a 3-D divergence operator should result naturally. Let $\mathbf{v}(x,y,z) = P(x,y,z)\mathbf{i} + Q(x,y,z)\mathbf{j} + R(x,y,z)\mathbf{k}$ for any 3-D vector field of interest. Considering

$$div\ \mathbf{v} \triangleq \frac{\partial P}{\partial x} + \frac{\partial Q}{\partial y} + \frac{\partial R}{\partial z}, \tag{4.67}$$

then we can define the 3-D divergence operator, as follows:

$$
\begin{aligned}
\text{DIV } \mathbf{v} \approx\ & \frac{P(i+1, j+\frac{1}{2}, k+\frac{1}{2}) - P(i, j+\frac{1}{2}, k+\frac{1}{2})}{(x_{i+1} - x_i)} + \\
& \frac{Q(i+\frac{1}{2}, j+1, k+\frac{1}{2}) - Q(i+\frac{1}{2}, j, k+\frac{1}{2})}{(y_{j+1} - y_j)} + \\
& \frac{R(i+\frac{1}{2}, j+\frac{1}{2}, k+1) - R(i+\frac{1}{2}, j+\frac{1}{2}, k)}{(z_{k+1} - z_k)}.
\end{aligned} \tag{4.68}
$$

It should be clear that the 3-D divergence is computed in the center

$$(x_{i+1/2}, y_{j+1/2}, z_{k+1/2}) \tag{4.69}$$

of a 3-D cell, based on the flux at the cell's faces.

4.4.3 3-D curl

As is made clear by the expression already found for the simplest CURL $\mathbf{v}(i, j+1/2)$, the evaluation of the 2-D CURL at some grid points will require values at the original nodes, as well as at the added nodes from staggering in the x and y directions.

Before exhibiting the 3-D staggering needed to evaluate *curl* $\mathbf{v}$ at some 3-D grid point, it is instructive to observe that, away from the boundaries, the Toeplitz-type structure of the first-order difference operators employed with CGM can be viewed as the result of using a **movable stencil for the grid points** laid along the 1-D staggered grid; a procedure that can be described as an x-**staggering**. It is also clear that, in order to maintain the desired order of accuracy, the distance between neighboring added points should be equal to the original grid spacing for **uniform** staggered grids.

Due to the relative complexity of the necessary 3-D staggering, we will now introduce a new notation for grid points; a notation motivated by the idea of describing the stencil of a point (x, y, z) **that is movable** throughout the 3-D fully staggered grid.

We use the following list of symbols (west, east, south, north, up, and down):

$$\blacktriangledown w = (x^-, y, z)\ ,\ \blacktriangledown e = (x^+, y, z), \tag{4.70}$$

$$\blacktriangledown s = (x, y^-, z)\ ,\ \blacktriangledown n = (x, y^+, z), \tag{4.71}$$

$$\blacktriangledown d = (x, y, z^-)\ ,\ \blacktriangledown u = (x, y, z^+), \tag{4.72}$$

For the 3-D vector field **v**, define the 3 auxiliary vectors:

$$\mathbf{v}_{xy}^* = \mathbf{v} \times \mathbf{k} = Q\mathbf{i} - P\mathbf{j} = P_{xy}^*\mathbf{i} + Q_{xy}^*\mathbf{j}\ ,\ P_{xy}^* = Q,\ Q_{xy}^* = -P, \quad (4.73)$$

$$\mathbf{v}_{yz}^* = \mathbf{v} \times \mathbf{i} = R\mathbf{j} - Q\mathbf{k} = Q_{yz}^*\mathbf{j} + R_{yz}^*\mathbf{k}\ ,\ Q_z^* = R,\ R_{yz}^* = -Q, \quad (4.74)$$

$$\mathbf{v}_{zx}^* = \mathbf{v} \times \mathbf{j} = P\mathbf{k} - R\mathbf{i} = R_{zx}^*\mathbf{k} + P_{zx}^*\mathbf{i}\ ,\ R_{zx}^* = P,\ P_{zx}^* = -R. \quad (4.75)$$

Let $\mathbf{x} = (x, y, z)$, we also have the following general expression for *curl* $\mathbf{v}(\mathbf{x})$:

$$curl\ v(\mathbf{x}) = \mathbf{i}\left(\frac{\partial R}{\partial y} - \frac{\partial Q}{\partial z}\right) + \mathbf{j}\left(\frac{\partial P}{\partial z} - \frac{\partial R}{\partial x}\right) + \mathbf{k}\left(\frac{\partial Q}{\partial x} - \frac{\partial P}{\partial y}\right) \quad (4.76)$$

$$= \mathbf{i}\ div\ \mathbf{v}_{yz}^* + \mathbf{j}\ div\ \mathbf{v}_{zx}^* + \mathbf{k}\ div\ \mathbf{v}_{xy}^*. \quad (4.77)$$

Therefore, the simplest approximation for *curl* $\mathbf{v}(x, y, z)$ is given at the **movable stencil center** (x, y, z), by

$$\text{CURL } \mathbf{v}(\mathbf{x}) = \mathbf{i}\left[\frac{Q_{yz}^*(\blacktriangledown n) - Q_{yz}^*(\blacktriangledown s)}{[y^+ - y^-]} + \frac{R_{yz}^*(\blacktriangledown u) - R_{yz}^*(\blacktriangledown d)}{[z^+ - z^-]}\right]$$

$$+ \mathbf{j}\left[\frac{R_{zx}^*(\blacktriangledown u) - R_{zx}^*(\blacktriangledown d)}{[z^+ - z^-]} + \frac{P_{zx}^*(\blacktriangledown e) - P_{zx}^*(\blacktriangledown w)}{[x^+ - x^-]}\right]$$

$$+ \mathbf{k}\left[\frac{P_{xy}^*(\blacktriangledown e) - P_{xy}^*(\blacktriangledown w)}{[x^+ - x^-]} + \frac{Q_{xy}^*(\blacktriangledown n) - Q_{xy}^*(\blacktriangledown s)}{[y^+ - y^-]}\right]. \quad (4.78)$$

The bracketed expressions correspond to the simplest numerical approximations for *div* $\mathbf{v}_{yz}^*$, *div* $\mathbf{v}_{zx}^*$, and *div* $\mathbf{v}_{xy}^*$ at the point (x, y, z). In this simple example, we need only the values of P_{xy}^*, Q_{xy}^*, Q_{yz}^*, R_{yz}^*, R_{zx}^*, and P_{zx}^* at the six stencil points $\blacktriangledown w$, $\blacktriangledown e$, $\blacktriangledown s$, $\blacktriangledown n$, $\blacktriangledown u$, and $\blacktriangledown d$, as previously described.

The 2-D divergences needed all arise from 2-D fluxes of vectors $\mathbf{v}_{yz}^*$, $\mathbf{v}_{zx}^*$ and $\mathbf{v}_{xy}^*$, and these fields lie in planes orthogonal to the coordinate axes.

As can be seen from Figure 4.9, $\mathbf{v}_{xy}^*$ lies in a plane orthogonal to the z-axis, and *div* $\mathbf{v}_{xy}^* = \mathbf{k} \cdot curl\ \mathbf{v}(x, y, z)$.

Analogously, $\mathbf{v}_{yz}^*$ lies in a plane orthogonal to the x-axis, and *div* $\mathbf{v}_{yz}^* = \mathbf{i} \cdot curl\ \mathbf{v}(x, y, z)$, and $\mathbf{v}_{zx}^*$ lies in a plane orthogonal to the y-axis, and *div* $\mathbf{v}_{zx}^* = \mathbf{j} \cdot curl\ \mathbf{v}(x, y, z)$.

The numerical approximation needs an individual staggering for each scalar component of *curl* $\mathbf{v}(x, y, z)$, in the way that is exhibited in Figure 4.9, but only for the z-component of *curl* $\mathbf{v}$. Naturally, for the full 3-D *curl* $\mathbf{v}(x, y, z)$, we need a combination of simultaneous staggerings in the x, y, and z-directions, and a rough idea of the way it looks, showing only some directional staggerings, is exhibited in Figure 4.10.

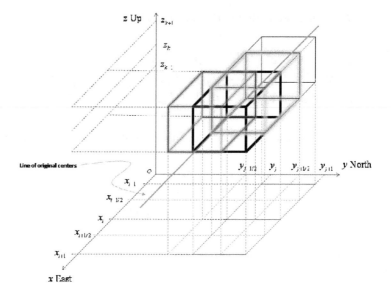

Figure 4.10: Partial 3-D staggering for 3-D CURL.

4.5   Gradient Compositions: $\nabla \cdot \nabla = \nabla^2$

For compositions involving the *grad* operator, such as the one occurring while dealing with Laplace's operator $div\ grad = \nabla^2$, it turns out that the gradient of some scalar function f will act as a vector-valued function upon which the *div* operator will have to act. This implies that *grad* f needs to be numerically computed at the centers of the faces of the original grid, which can be accomplished by means of an additional auxiliary staggering, thus resulting in a full 2-D staggered grid as shown in Figure 4.11 (the auxiliary staggered grid is depicted in blue). The computed Castillo–Grone Laplacian is naturally defined as $\check{\mathbf{L}} = \mathring{\mathbf{D}}\check{\mathbf{G}}$.

In Figure 4.11, we observe that a nabla symbol has been used to allude to the location of the computed Laplacian, which is the cell center of the original 2-D grid; also noteworthy is the fact that the coordinates of the cell centers are both described by fractional indexes. In order to compute the required Laplacian, the following two collections of values are needed: First, the collection of computed gradients at points along the x-direction; and second, the collection of computed gradients at points along the y-direction. For the first collection we use red, x-oriented diamonds to indicate the location of the points at which the gradient is computed using blue, y-oriented diamonds for the second collection.

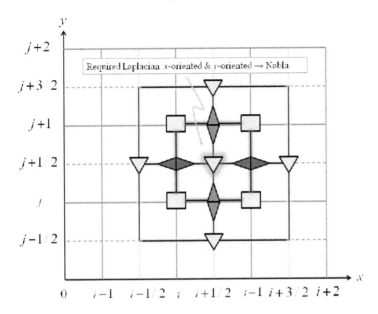

Figure 4.11: Full 2-D staggered grid for compositions with GRAD.

For the sake of the future 3-D extension, it is important to understand that diamond-denoted nodes, regardless of their orientation, are always hybrid nodes, as defined in the following section.

4.5.1 Hybrid Nodes and Computing the Laplacian

In previous sections (Figure 4.6), we stated that diamond-denoted nodes can be thought of as **hybrid nodes**. In this section, we present a few noteworthy remarks on these nodes.

In the case of hybrid nodes, if one of their coordinates belongs to the nodal grid, then the other coordinate must belong to the set of added staggered nodes. These can then be depicted, while simultaneously attempting to reach the full 2-D staggering (in order to compute the Laplacian), by computing the full gradient at the x-oriented hybrid nodes; and then, to compute the gradient in the y-hybrid nodes, we determine that the latter ones are merely duals of the first ones (Figure 4.12). Figure 4.13 depicts this for the 3-D case.

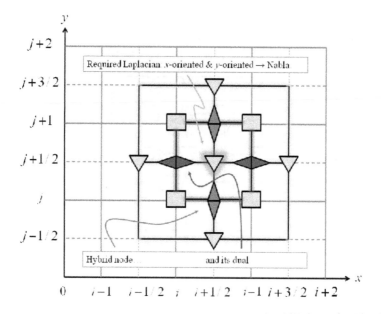

Figure 4.12: Full 2-D staggered grid for compositions with GRAD depicting the concepts of hybrid and dual nodes.

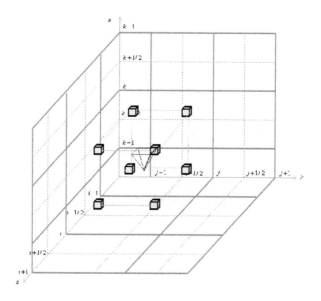

Figure 4.13: A 3-D gradient composition yielding a 3-D Laplacian at a cell center.

4.6 Nullity Tests

As mentioned in §4.1, it is desired that the constructed discrete analogs **D**, **G**, and **C** of *div*, *grad*, and *curl* continuous operators, respectively, should satisfy the nullity conditions for the compositions of the operators listed in Table 5.1; that is:

$$\mathbf{CG}f = 0,$$
$$\mathbf{DCv} = 0.$$

In the continuous case, the corresponding nullity properties (in all cases) are a consequence of Schwarz' theorem, about the equality of mixed partial derivatives. This fact leads us to design the following simple test for the analog property for discrete first-order derivatives:

$$\mathbf{D}_2(\mathbf{i}(\mathbf{G}f \cdot \mathbf{j})) = \mathbf{D}_2(\mathbf{j}(\mathbf{G}f \cdot \mathbf{i})), \tag{4.79}$$

where $\mathbf{D}_2$ stands for our 2-D DIV.

4.7 Higher-Order Operators

4.7.1 Sixth-Order Operators

In the previous section, we introduced second- and fourth-order operators. In this section, we present operators yielding sixth-order accurate solutions, uniformly across the domain of interest.

As for the case of second- and fourth-order operators, it is possible to construct **families** of uniformly sixth- and eighth-order accurate gradients and divergences (see [148]). For example, take the following uniformly sixth-order accurate gradient $h\mathbf{G}$ (here, we present only the number of rows necessary to illustrate the overall structure of the operators):

$$
\begin{bmatrix}
g_{11} & g_{12} & g_{13} & g_{14} & g_{15} & g_{16} & g_{17} & g_{18} & g_{19} & 0 \cdots \\
g_{21} & g_{22} & g_{23} & g_{24} & g_{25} & g_{26} & g_{27} & g_{28} & g_{29} & 0 \cdots \\
g_{31} & g_{32} & g_{33} & g_{34} & g_{35} & g_{36} & g_{37} & g_{38} & g_{39} & 0 \cdots \\
-\frac{9}{1920} & \frac{125}{1920} & -\frac{2250}{1920} & \frac{2250}{1920} & -\frac{125}{1920} & \frac{9}{1920} & 0 & 0 & 0 & 0 \cdots \\
0 & -\frac{9}{1920} & \frac{125}{1920} & -\frac{2250}{1920} & \frac{2250}{1920} & -\frac{125}{1920} & \frac{9}{1920} & 0 & 0 & 0 \cdots \\
0 & 0 & -\frac{9}{1920} & \frac{125}{1920} & -\frac{2250}{1920} & \frac{2250}{1920} & -\frac{125}{1920} & \frac{9}{1920} & 0 & 0 \cdots \\
0 & 0 & 0 & -\frac{9}{1920} & \frac{125}{1920} & -\frac{2250}{1920} & \frac{2250}{1920} & -\frac{125}{1920} & \frac{9}{1920} & 0 \cdots
\end{bmatrix}, \tag{4.80}
$$

where, for the first row, we have

$$g_{11} = -\frac{568557184}{150834915} \quad g_{12} = \frac{455704609}{835795520} \quad g_{13} = -\frac{128942179}{417897760}$$

$$g_{14} = \frac{15911389}{6964960} \quad g_{15} = -\frac{142924471}{117011328} \quad g_{16} = \frac{20331719}{50147712}$$

$$g_{17} = -\frac{2688571}{38307280} \quad g_{18} = \frac{187529}{41789760} \quad g_{19} = -\frac{6207}{27859840}.$$

Similarly, for the second row, we have

$$g_{21} = \frac{496}{3465} \quad g_{22} = -\frac{811}{640} \quad g_{23} = \frac{449}{384}$$

$$g_{24} = -\frac{29}{960} \quad g_{25} = -\frac{11}{448} \quad g_{26} = \frac{13}{112}$$

$$g_{27} = -\frac{37}{21120} \quad g_{28} = 0 \quad g_{29} = 0.$$

Finally, for the third row, we have

$$g_{31} = -\frac{8}{385} \quad g_{32} = \frac{179}{1920} \quad g_{33} = -\frac{153}{128}$$

$$g_{34} = \frac{381}{320} \quad g_{35} = -\frac{101}{1344} \quad g_{36} = \frac{1}{128}$$

$$g_{37} = -\frac{3}{7040} \quad g_{38} = 0 \quad g_{39} = 0.$$

For the latter operator, the related weight matrix P is defined, as follows:

$$P = \begin{bmatrix} p_{11} & 0 & 0 & 0 & 0 & 0 & 0 & 0 & \cdots & & \cdots & & 0 \\ 0 & p_{22} & 0 & 0 & 0 & 0 & 0 & 0 & \cdots & & \cdots & & 0 \\ 0 & 0 & p_{33} & 0 & 0 & 0 & 0 & 0 & \cdots & & \cdots & & 0 \\ 0 & 0 & 0 & p_{44} & 0 & 0 & 0 & 0 & \cdots & & \cdots & & 0 \\ 0 & 0 & 0 & 0 & p_{55} & 0 & 0 & 0 & \cdots & & \cdots & & 0 \\ 0 & 0 & 0 & 0 & 0 & p_{66} & 0 & 0 & \cdots & & \cdots & & 0 \\ 0 & 0 & 0 & 0 & 0 & 0 & 1 & 0 & \cdots & & \cdots & & 0 \\ \vdots & \cdots & & & & \cdots & & \ddots & \cdots & & \cdots & & \vdots \\ 0 & \cdots & & & \cdots & 0 & 1 & 0 & 0 & 0 & 0 & 0 & 0 \\ 0 & \cdots & & & \cdots & 0 & 0 & p_{66} & 0 & 0 & 0 & 0 & 0 \\ 0 & \cdots & & & \cdots & 0 & 0 & 0 & p_{55} & 0 & 0 & 0 & 0 \\ 0 & \cdots & & & \cdots & 0 & 0 & 0 & 0 & p_{44} & 0 & 0 & 0 \\ 0 & \cdots & & & \cdots & 0 & 0 & 0 & 0 & 0 & p_{33} & 0 & 0 \\ 0 & \cdots & & & \cdots & 0 & 0 & 0 & 0 & 0 & 0 & p_{22} & 0 \\ 0 & \cdots & & & \cdots & 0 & 0 & 0 & 0 & 0 & 0 & 0 & p_{11} \end{bmatrix}, \quad (4.81)$$

where

$$p_{11} = \frac{43531}{138240} \quad p_{22} = \frac{192937}{138240} \quad p_{33} = \frac{42647}{69120}$$

$$p_{44} = \frac{86473}{69120} \quad p_{55} = \frac{125303}{138240} \quad p_{66} = \frac{140309}{138240}.$$

Analogously, we introduce the following uniformly sixth-order accurate divergence $h\check{\mathbf{D}}$:

$$\begin{bmatrix} d_{11} & d_{12} & d_{13} & d_{14} & d_{15} & d_{16} & d_{17} & d_{18} & d_{19} & 0 & \cdots \\ d_{21} & d_{22} & d_{23} & d_{24} & d_{25} & d_{26} & d_{27} & d_{28} & d_{29} & 0 & \cdots \\ -\frac{9}{1920} & \frac{125}{1920} & -\frac{2250}{1920} & \frac{2250}{1920} & -\frac{125}{1920} & \frac{9}{1920} & 0 & 0 & 0 & 0 & \cdots \\ 0 & -\frac{9}{1920} & \frac{125}{1920} & -\frac{2250}{1920} & \frac{2250}{1920} & -\frac{125}{1920} & \frac{9}{1920} & 0 & 0 & 0 & \cdots \\ 0 & 0 & -\frac{9}{1920} & \frac{125}{1920} & -\frac{2250}{1920} & \frac{2250}{1920} & -\frac{125}{1920} & \frac{9}{1920} & 0 & 0 & \cdots \\ 0 & 0 & 0 & -\frac{9}{1920} & \frac{125}{1920} & -\frac{2250}{1920} & \frac{2250}{1920} & -\frac{125}{1920} & \frac{9}{1920} & 0 & \cdots \end{bmatrix}, \quad (4.82)$$

where, for the first row, we have

$$d_{11} = -\frac{1077397}{1273920} \quad d_{12} = \frac{15668474643803}{32472850116480} \quad d_{13} = \frac{49955527}{39491520}$$

$$d_{14} = -\frac{25369793}{19745760} \quad d_{15} = \frac{12220145}{15796608} \quad d_{16} = -\frac{21334421}{78983040}$$

$$d_{17} = \frac{460217}{9872880} \quad d_{18} = -\frac{101017}{39491520} \quad d_{19} = \frac{3369}{26327680}.$$

For the second row, we have

$$d_{21} = \frac{31}{960} \quad d_{22} = -\frac{687}{640} \quad d_{23} = \frac{129}{128}$$

$$d_{24} = \frac{19}{192} \quad d_{25} = -\frac{3}{32} \quad d_{26} = \frac{21}{640}$$

$$d_{27} = -\frac{3}{640} \quad d_{28} = 0 \quad d_{29} = 0.$$

For the latter operator, the related weight matrix Q is defined, as follows:

$$Q = \begin{bmatrix} q_{11} & 0 & 0 & 0 & 0 & 0 & 0 & 0 & \cdots & & \cdots & 0 \\ 0 & q_{22} & 0 & 0 & 0 & 0 & 0 & 0 & \cdots & & \cdots & 0 \\ 0 & 0 & q_{33} & 0 & 0 & 0 & 0 & 0 & \cdots & & \cdots & 0 \\ 0 & 0 & 0 & q_{44} & 0 & 0 & 0 & 0 & \cdots & & \cdots & 0 \\ 0 & 0 & 0 & 0 & q_{55} & 0 & 0 & 0 & \cdots & & \cdots & 0 \\ 0 & 0 & 0 & 0 & 0 & q_{66} & 0 & 0 & \cdots & & \cdots & 0 \\ 0 & 0 & 0 & 0 & 0 & 0 & 1 & 0 & \cdots & & \cdots & 0 \\ \vdots & \cdots & & & & \cdots & & \ddots & \cdots & & & \vdots \\ 0 & \cdots & & & & \cdots & 0 & 1 & 0 & 0 & 0 & 0 & 0 & 0 \\ 0 & \cdots & & & & \cdots & 0 & 0 & q_{66} & 0 & 0 & 0 & 0 & 0 \\ 0 & \cdots & & & & \cdots & 0 & 0 & 0 & q_{55} & 0 & 0 & 0 & 0 \\ 0 & \cdots & & & & \cdots & 0 & 0 & 0 & 0 & q_{44} & 0 & 0 & 0 \\ 0 & \cdots & & & & \cdots & 0 & 0 & 0 & 0 & 0 & q_{33} & 0 & 0 \\ 0 & \cdots & & & & \cdots & 0 & 0 & 0 & 0 & 0 & 0 & q_{22} & 0 \\ 0 & \cdots & & & & \cdots & 0 & 0 & 0 & 0 & 0 & 0 & 0 & q_{11} \end{bmatrix}, \quad (4.83)$$

where

$$q_{11} = \frac{41137}{34560} \quad q_{22} = \frac{15667}{34560} \quad q_{33} = \frac{2933}{1728}$$

$$q_{44} = \frac{2131}{4320} \quad q_{55} = \frac{41411}{34560} \quad q_{66} = \frac{33437}{34560}.$$

4.7.2 Eighth-Order Operators

In this section, we present operators yielding eighth-order accurate solutions. Recall that we present only the number of rows necessary to illustrate the overall structure of the operators. First, we introduce the uniformly eighth-order accurate gradient: $h\check{\mathbf{G}}$:

$$
\begin{bmatrix}
g_{11} & g_{12} & g_{13} & g_{14} & g_{15} & g_{16} & g_{17} & g_{18} & g_{19} & \cdots & g_{112} & 0 & \cdots \\
g_{21} & g_{22} & g_{23} & g_{24} & g_{25} & g_{26} & g_{27} & g_{28} & g_{29} & \cdots & g_{212} & 0 & \cdots \\
g_{31} & g_{32} & g_{33} & g_{34} & g_{35} & g_{36} & g_{37} & g_{38} & g_{39} & \cdots & g_{312} & 0 & \cdots \\
g_{41} & g_{42} & g_{43} & g_{44} & g_{45} & g_{46} & g_{47} & g_{48} & g_{49} & \cdots & g_{412} & 0 & \cdots \\
\frac{5}{7168} & -\frac{49}{5120} & \frac{245}{3072} & -\frac{1225}{1024} & \frac{1225}{1024} & -\frac{245}{3072} & \frac{49}{5120} & -\frac{5}{7168} & 0 & & 0 & \cdots & 0 & \cdots \\
0 & \frac{5}{7168} & -\frac{49}{5120} & \frac{245}{3072} & -\frac{1225}{1024} & \frac{1225}{1024} & -\frac{245}{3072} & \frac{49}{5120} & -\frac{5}{7168} & 0 & & 0 & \cdots & 0 & \cdots
\end{bmatrix},
$$

$$(4.84)$$

where, for the first row, we have

$$
\begin{aligned}
g_{11} &= -\frac{375430666840256}{92579164853175} & g_{12} &= \frac{46577871283831}{7366050101760} & g_{13} &= -\frac{44142164823881}{8839260122112} \\
g_{14} &= \frac{50703079390921}{9207562627200} & g_{15} &= -\frac{71068924474957}{14732100203520} & g_{16} &= \frac{57866887554917}{18941271690240} \\
g_{17} &= -\frac{30717060475411}{23150443176960} & g_{18} &= \frac{2027314948429}{5471922932736} & g_{19} &= -\frac{13407250027393}{220981503052800} \\
g_{110} &= \frac{82765484227}{14732100203520} & g_{111} &= -\frac{1177332481}{2455350033920} & g_{113} &= \frac{21454295}{982140013568}.
\end{aligned}
$$

Similarly, for the second row, we have

$$
\begin{aligned}
g_{21} &= \frac{86048}{675675} & g_{22} &= -\frac{131093}{107520} & g_{23} &= \frac{49087}{46080} \\
g_{24} &= \frac{10973}{76800} & g_{25} &= -\frac{4597}{21504} & g_{26} &= \frac{4019}{27648} \\
g_{27} &= -\frac{10331}{168960} & g_{28} &= \frac{2983}{199680} & g_{29} &= -\frac{2621}{1612800} \\
g_{210} &= 0 & g_{211} &= 0 & g_{212} &= 0.
\end{aligned}
$$

For the third row, we have

$$
\begin{aligned}
g_{31} &= -\frac{3776}{225225} & g_{32} &= \frac{8707}{107520} & g_{33} &= -\frac{17947}{15360} \\
g_{34} &= \frac{29319}{25600} & g_{35} &= -\frac{533}{21504} & g_{36} &= -\frac{263}{9216} \\
g_{37} &= \frac{903}{56320} & g_{38} &= -\frac{283}{66560} & g_{39} &= \frac{257}{537600} \\
g_{310} &= 0 & g_{311} &= 0 & g_{312} &= 0.
\end{aligned}
$$

Finally, for the fourth row, we have

$$
\begin{aligned}
g_{41} &= \frac{32}{9009} & g_{42} &= -\frac{543}{35840} & g_{43} &= \frac{265}{3072} \\
g_{44} &= -\frac{1233}{1024} & g_{45} &= \frac{8625}{7168} & g_{46} &= -\frac{775}{9216} \\
g_{47} &= \frac{639}{56320} & g_{48} &= -\frac{15}{13312} & g_{49} &= \frac{1}{21504} \\
g_{410} &= 0 & g_{411} &= 0 & g_{412} &= 0.
\end{aligned}
$$

For the latter operator, the related weight matrix P is defined as the following matrix:

$$
\begin{bmatrix}
p_{11} & 0 & 0 & 0 & 0 & 0 & 0 & 0 & 0 & 0 & \cdots & & & \cdots & 0 \\
0 & p_{22} & 0 & 0 & 0 & 0 & 0 & 0 & 0 & 0 & \cdots & & & \cdots & 0 \\
0 & 0 & p_{33} & 0 & 0 & 0 & 0 & 0 & 0 & 0 & \cdots & & & \cdots & 0 \\
0 & 0 & 0 & p_{44} & 0 & 0 & 0 & 0 & 0 & 0 & \cdots & & & \cdots & 0 \\
0 & 0 & 0 & 0 & p_{55} & 0 & 0 & 0 & 0 & 0 & \cdots & & & \cdots & 0 \\
0 & 0 & 0 & 0 & 0 & p_{66} & 0 & 0 & 0 & 0 & \cdots & & & \cdots & 0 \\
0 & 0 & 0 & 0 & 0 & 0 & p_{77} & 0 & 0 & 0 & \cdots & & & \cdots & 0 \\
0 & 0 & 0 & 0 & 0 & 0 & 0 & p_{88} & 0 & 0 & \cdots & & & \cdots & 0 \\
0 & 0 & 0 & 0 & 0 & 0 & 0 & 0 & 1 & 0 & \cdots & & & \cdots & 0 \\
\vdots & & \cdots & & & & \cdots & & \ddots & & \cdots & & & \cdots & \vdots \\
0 & \cdots & & & & & \cdots & 0 & 1 & 0 & 1 & 0 & 0 & 0 & 0 & 0 & 0 \\
0 & \cdots & & & & & \cdots & 0 & 0 & p_{88} & 0 & 0 & 0 & 0 & 0 & 0 & 0 \\
0 & \cdots & & & & & \cdots & 0 & 0 & 0 & p_{77} & 0 & 0 & 0 & 0 & 0 & 0 \\
0 & \cdots & & & & & \cdots & 0 & 0 & 0 & 0 & p_{66} & 0 & 0 & 0 & 0 & 0 \\
0 & \cdots & & & & & \cdots & 0 & 0 & 0 & 0 & 0 & p_{55} & 0 & 0 & 0 & 0 \\
0 & \cdots & & & & & \cdots & 0 & 0 & 0 & 0 & 0 & 0 & p_{44} & 0 & 0 & 0 \\
0 & \cdots & & & & & \cdots & 0 & 0 & 0 & 0 & 0 & 0 & 0 & p_{33} & 0 & 0 \\
0 & \cdots & & & & & \cdots & 0 & 0 & 0 & 0 & 0 & 0 & 0 & 0 & p_{22} & 0 \\
0 & \cdots & & & & & \cdots & 0 & 0 & 0 & 0 & 0 & 0 & 0 & 0 & 0 & p_{11}
\end{bmatrix},
\tag{4.85}
$$

where

$$
\begin{aligned}
& p_{11} = \tfrac{137017301}{464486400} && p_{22} = \tfrac{708364333}{464486400} && p_{33} = \tfrac{13391089}{51609600} \\
& p_{44} = \tfrac{33345545}{18579456} && p_{55} = \tfrac{38651603}{92897280} && p_{66} = \tfrac{65878031}{51609600} \\
& p_{77} = \tfrac{429170387}{464486400} && p_{88} = \tfrac{468777259}{464486400}.
\end{aligned}
$$

Analogously, we introduce the following uniformly eighth-order accurate divergence $h\breve{\mathbf{D}}$: $h\breve{\mathbf{G}}$:

$$
\begin{bmatrix}
d_{11} & d_{12} & d_{13} & d_{14} & d_{15} & d_{16} & d_{17} & d_{18} & d_{19} & \cdots & d_{112} & 0 & \cdots \\
d_{21} & d_{22} & d_{23} & d_{24} & d_{25} & d_{26} & d_{27} & d_{28} & d_{29} & \cdots & d_{212} & 0 & \cdots \\
d_{31} & d_{32} & d_{33} & d_{34} & d_{35} & d_{36} & d_{37} & d_{38} & d_{39} & \cdots & d_{312} & 0 & \cdots \\
\tfrac{5}{7168} & -\tfrac{49}{5120} & \tfrac{245}{3072} & -\tfrac{1225}{1024} & \tfrac{1225}{1024} & -\tfrac{245}{3072} & \tfrac{49}{5120} & -\tfrac{5}{7168} & 0 & 0 & \cdots & 0 & \cdots \\
0 & \tfrac{5}{7168} & -\tfrac{49}{5120} & \tfrac{245}{3072} & -\tfrac{1225}{1024} & \tfrac{1225}{1024} & -\tfrac{245}{3072} & \tfrac{49}{5120} & -\tfrac{5}{7168} & 0 & \cdots & 0 & \cdots
\end{bmatrix},
\tag{4.86}
$$

where, for the first row, we have

$$
\begin{aligned}
& d_{11} = -\tfrac{12379364146687}{15622313710080} && d_{12} = -\tfrac{436022799711}{5207437903360} && d_{13} = \tfrac{80790752734709}{31244627420160} \\
& d_{14} = -\tfrac{4898774991147}{1301859475840} && d_{15} = \tfrac{3484785616723}{946806891520} && d_{16} = -\tfrac{3665347685293}{1487839400960} \\
& d_{17} = \tfrac{1635418471121}{1487839400960} && d_{18} = -\tfrac{3208052016403}{10414875806720} && d_{19} = \tfrac{1527953922703}{31244627420160} \\
& d_{110} = -\tfrac{129082472809}{31244627420160} && d_{111} = \tfrac{1849044667}{5207437903360} && d_{113} = -\tfrac{33958565}{2082975161344}.
\end{aligned}
$$

Similarly, for the second row, we have

$$d_{21} = \frac{2689}{107520} \quad d_{22} = -\frac{36527}{35840} \quad d_{23} = \frac{4259}{5120}$$

$$d_{24} = \frac{6497}{15360} \quad d_{25} = -\frac{475}{1024} \quad d_{26} = \frac{1541}{5120}$$

$$d_{27} = -\frac{639}{5120} \quad d_{28} = \frac{1087}{35840} \quad d_{29} = -\frac{59}{17920}$$

$$d_{210} = 0 \quad d_{211} = 0 \quad d_{212} = 0.$$

Finally, for the third row, we have

$$d_{31} = -\frac{59}{17920} \quad d_{32} = \frac{1175}{21504} \quad d_{33} = -\frac{1165}{1024}$$

$$d_{34} = \frac{1135}{1024} \quad d_{35} = \frac{25}{3072} \quad d_{36} = -\frac{251}{5120}$$

$$d_{37} = \frac{25}{1024} \quad d_{38} = -\frac{45}{7168} \quad d_{39} = \frac{5}{7168}$$

$$d_{310} = 0 \quad d_{311} = 0 \quad d_{312} = 0.$$

For the latter operator, the related weight matrix Q is defined, as follows:

$$
\begin{bmatrix}
q_{11} & 0 & 0 & 0 & 0 & 0 & 0 & 0 & 0 & 0 & \cdots & & \cdots & 0 \\
0 & q_{22} & 0 & 0 & 0 & 0 & 0 & 0 & 0 & 0 & \cdots & & \cdots & 0 \\
0 & 0 & q_{33} & 0 & 0 & 0 & 0 & 0 & 0 & 0 & \cdots & & \cdots & 0 \\
0 & 0 & 0 & q_{44} & 0 & 0 & 0 & 0 & 0 & 0 & \cdots & & \cdots & 0 \\
0 & 0 & 0 & 0 & q_{55} & 0 & 0 & 0 & 0 & 0 & \cdots & & \cdots & 0 \\
0 & 0 & 0 & 0 & 0 & q_{66} & 0 & 0 & 0 & 0 & \cdots & & \cdots & 0 \\
0 & 0 & 0 & 0 & 0 & 0 & q_{77} & 0 & 0 & 0 & \cdots & & \cdots & 0 \\
0 & 0 & 0 & 0 & 0 & 0 & 0 & q_{88} & 0 & 0 & \cdots & & \cdots & 0 \\
0 & 0 & 0 & 0 & 0 & 0 & 0 & 0 & 1 & 0 & \cdots & & \cdots & 0 \\
\vdots & \cdots & & & & & \cdots & & \ddots & \cdots & & & \cdots & \vdots \\
0 & \cdots & & & & & \cdots & 0 & 1 & 0 & 1 & 0 & 0 & 0 & 0 & 0 & 0 \\
0 & \cdots & & & & & \cdots & 0 & 0 & q_{88} & 0 & 0 & 0 & 0 & 0 & 0 & 0 \\
0 & \cdots & & & & & \cdots & 0 & 0 & 0 & q_{77} & 0 & 0 & 0 & 0 & 0 & 0 \\
0 & \cdots & & & & & \cdots & 0 & 0 & 0 & 0 & q_{66} & 0 & 0 & 0 & 0 & 0 \\
0 & \cdots & & & & & \cdots & 0 & 0 & 0 & 0 & 0 & q_{55} & 0 & 0 & 0 & 0 \\
0 & \cdots & & & & & \cdots & 0 & 0 & 0 & 0 & 0 & 0 & q_{44} & 0 & 0 & 0 \\
0 & \cdots & & & & & \cdots & 0 & 0 & 0 & 0 & 0 & 0 & 0 & q_{33} & 0 & 0 \\
0 & \cdots & & & & & \cdots & 0 & 0 & 0 & 0 & 0 & 0 & 0 & 0 & q_{22} & 0 \\
0 & \cdots & & & & & \cdots & 0 & 0 & 0 & 0 & 0 & 0 & 0 & 0 & 0 & q_{11}
\end{bmatrix},
$$
(4.87)

where

$$q_{11} = \frac{290593633}{232243200} \quad q_{22} = \frac{13734569}{232243200} \quad q_{33} = \frac{71825597}{25804800}$$

$$q_{44} = -\frac{7678657}{6635520} \quad q_{55} = \frac{24991643}{9289728} \quad q_{66} = \frac{4301443}{25804800}$$

$$q_{77} = \frac{286984471}{232243200} \quad q_{88} = \frac{225451487}{232243200}.$$

4.8 Formulation of Nonlinear and Time-Dependent Problems

An example of a time-dependent problem, in which mimetic discretization methods have been used, can be found in [149]. The problem can be formulated as follows. Consider the following equation:

$$\frac{\partial u}{\partial t} - \frac{\partial^2 u}{\partial x^2} = -\frac{1}{10}e^{-\frac{t}{10}}\sin(2\pi x) + 4e^{-\frac{t}{10}}\sin(2\pi x)\pi^2, \qquad (4.88)$$

defined in $(0,1)$. Robin's conditions are specified, as follows:

$$u(0) - \frac{\partial u}{\partial x}(0) = -2\pi e^{-\frac{t}{10}} \qquad (4.89)$$

$$u(1) + \frac{\partial u}{\partial x}(1) = 2\pi e^{-\frac{t}{10}}, \qquad (4.90)$$

with the following initial condition:

$$u(x,0) = e^{\frac{t}{10}}\sin(2\pi x). \qquad (4.91)$$

A Crank–Nicholson finite-difference scheme was used to discretize the time-derivative. In [149], the following analytic solution was considered in order to verify the validity of the numerical solutions:

$$u(x,0) = \sin(2\pi x). \qquad (4.92)$$

Nonlinear problems can be examined in a similar manner. For example, consider Burger's equations for inviscid flow:

$$u_t + u_x = 0 \qquad (4.93)$$

$$u_t + \left(\frac{1}{2}u^2\right)_x = 0. \qquad (4.94)$$

The following initial condition can be included in the formulation:

$$u(x,0) = \exp(-10(4x-1)^2). \qquad (4.95)$$

Section 8.5 expands on the application of mimetic discretization methods in fluid dynamics.

4.9 Concluding Remarks

In this chapter we introduced the most important concepts related to mimetic operators. We presented the Castillo–Grone operators and explained their characteristics regarding their ability to mimic the most valuable properties of their continuum counterparts. We also examined the fundamental considerations for extending their usage to higher-dimensional scenarios, and presented higher-order operators and their uses (particularly, fourth- and sixth-order operators).

Additionally, we illustrated some of the most important considerations which arise when defining gradient compositions; as in the case of the Laplacian, as well as explaining a nullity test that can be considered as a means of verifying the mimetic nature of these operators.

4.10   Sample Problems

1. Write code to solve the Poisson equation $\Delta u = f$ on a square $m \times m$ grid with $\Delta x = \Delta y = h$, using the FD method known as the 5-point Laplacian. It is set up to solve a test problem, for which the exact solution is $u(x, y) = \exp(x + y/2)$, using Dirichlet boundary conditions and the right-hand side $f(x, y) = 1.25 \exp(x + y/2)$.

   (a) Test this code by performing a grid refinement study to verify that it is second-order accurate. A typical table for comparing $(\hat{\mathbf{D}}\check{\mathbf{G}}\tilde{u} - \Delta u)$ at a given grid point might involve a grid refinement defined by the following sequence of h-values:

      1.0e-01, 5.0e-02, 1.0e-02, 5.0e-03 and 1.0e-03.

   (b) Modify the code so that it works on a rectangular domain $[a_x, b_x] \times [a_y, b_y]$, but still with $\Delta x = \Delta y = h$. Test your modified code with some specific non-square domain. Make sure you validate the feasibility of the selected definitory values for the rectangular domain with respect to the selected common value for h.

2. Create a different code so that it uses the Castillo–Grone Laplacian defined in §3.5.1 as $\check{\mathbf{L}} = \hat{\mathbf{D}}\check{\mathbf{G}}$.

   (a) For the 1-D uniform case, compare the solution by means of FD and a mimetic method.

3. Verify that the proposed simple nullity tests in §3.6 become the discrete analog for the equality of mixed partial derivatives, when the second-order accurate Castillo–Grone operator $\check{\mathbf{G}}$ and the resulting D_2 are used.

4. Propose an algorithm for the implementation of the Castillo–Grone 2-D divergence and 2-D gradient. Can you compute the Castillo–Grone Laplacian? Can you propose a problem in which these are of utility? What about higher-order operators?

5. Propose an algorithm for the implementation of the Castillo–Grone 3-D divergence and 3-D gradient. Can you compute the Castillo–Grone Laplacian? Can you propose a problem in which these are of utility? What about higher-order operators?

Chapter 5

Object-Oriented Programming and C++

In this chapter, we present some of the most important concepts in object-oriented programming, as a means of providing a thorough theoretical background for the concepts we will examine in Chapter 6. We explain the basics of object orientation as a programming paradigm, as well as the fundamentals of how C++ implements this paradigm. Our intention is to present the most essential concepts, while depicting their applications for numerically solving problems in Computational Science. In summary: Our main objective is to intuitively introduce the reader to the concept of an object-oriented application programming interface (API), as it will be considered in Chapter 6.

5.1 From Structured to Object-Oriented Programming

It is widely known that computer programming's theoretical foundations are based on mathematical logic and algorithms. However, although the theoretical foundations may be the same, there are a variety of programming methodologies in existence, since problems in science are necessarily considered from a diverse range of perspectives. We call each different programming methodology a **programming paradigm**.

Structured programming is one of the most widely known programming paradigms. In structured programming, algorithms are conceptualized as a finite sequence of instructions that eventually yield the desired solution to a given problem of interest. This sequence of instructions is generally non-commutative, and should be free of ambiguities. Structured programming, first made its appearance in the 1960s, as a response to the emerging complexities of the problems then under consideration. An important work on the subject is a famous letter written by the eminent computer science theorist, Edsger W. Dijkstra [150].

Structured programming makes use of the concept of **modules** in order to conceptualize a problem as a finite sequence of instructions, among which, problems of a less complex nature can be studied and solved more easily. Modules represent a valuable modeling resource, since they can be thought of as specific operations performed by certain entities that arise from the given problem under consideration. In structured programming, these entities are implemented by means of **structures**. A structure is a collection of conceptually related data, possessing its own meaning, given the nature of the modeled scenario. Together, a given structure and its collection of operations, which are implemented as modules, are called an **abstract data type (ADT)**. This name was motivated by the fact that programming languages (which can implement one or several programming paradigms) possess their own native data types; thereby, allowing the programmer to construct the programs by means of the control mechanisms that the various languages provide. Therefore, ADTs should be defined in terms of native data types, or, given their complexity, in terms of other ADTs, which can eventually be narrowed down and defined in terms of native data types.

One example of an ADT is a data type which allows the programmer the use of points defined within a 3-D space. If we name the ADT `Point3D`, then we can define the following structure for its representation:

```
struct Point3D{
    double xx_;
    double yy_;
    double zz_;
    double norm2_;
    int index_;
};
```

In the previous snippet of code, we see that a point in a 3-D space can be conceptualized as a collection of five related **data fields**; which data fields having been implemented by native data types. Specifically, we have decided that the information on the three coordinates defining the point is important, so we have decided to implement them as double precision floating point numbers. Similarly, we have decided that the value of the Euclidean norm for the point is also important, since it can be easily defined in terms of the given coordinates. Finally, we have chosen to identify each point using an integer index value. Notice the conventions used when naming both the structure and its fields.

To complete the ADT, we need operations to interact with our created entity. One basic operation we can use is the creation of a default instance of a 3-D point, which consists of ensuring all of the fields are initialized with their default value, given the data type they have been implemented as. In this case, the default value is zero. We can also specify the previously known values for all of the fields, or just for the coordinates and the index, while letting the operation compute the norm, as follows:

```
bool Create(Point3D *in, double xx, double yy, double zz, int
    index) {
    /* Validate given data: */
    if (in == NULL) {
```

```
 4      return false;
 5    }
 6    if (index <= 0) {
 7      return false;
 8    }
 9    /* Define the Point3D in terms of the given values: */
10    in->xx_ = xx;
11    in->yy_ = yy;
12    in->zz_ = zz;
13    in->index_ = index;
14    /* The operation takes care of computing the norm: */
15    in->norm2_ = sqrt(xx*xx + yy*yy + zz*zz);
16    return true;
17  }
```

Notice the utility of the adopted conventions when naming elements in the code.

In the previous snippet of code, we guaranteed the validity of the created instance by means of having an operation to take care of it. Our defined operation ensured that correct data were provided, and that they were handled correctly. Similar operations could also have been developed that would have been made responsible for the validity and consistency of the ADT instances they are related to. The practice of "hiding" the internal details of an ADT so that only a finite set of operations can interact with them, is called **implementation hiding**.

As the complexity of the considered problems kept increasing, systems required extra features in order to be efficiently modeled in computer programming. One example is the necessity of also handling 2-D points in a certain application. Structured programming would require the creation of a new structure, namely **2DPoint**, to achieve this purpose, thus forcing the programmer to practically rewrite the code for the 3-D case, since it can be easily seen that the 2-D case is a specific instance of the 3-D case. Furthermore, implementation hiding is not supported by structured programming; therefore, programmers were prone to programming errors, given the potential for the involuntary modification of a data field within a given instance of an ADT.

Naturally, programmers developed object-oriented programming, as a response to these, and other problems they were facing.

In **object-oriented programming (OOP)**, structures are implemented as **classes**, which are analogous to structures, except that they support more mechanisms to ensure their own consistency and implementation hiding. An instance of a class is called an **object**, hence, the name of the paradigm.

5.2   Fundamental Concepts in Object-Oriented Programming

In order to introduce the most important concepts in OOP, we will present an example in the context of numerically solving a PDE. Specifically, we will study how to represent a uniform grid by means of a class. We will name our class `LogicallyRectangular1DUniformNodalGrid`. There is an excellent reason for such a long name: We want descriptive names for what we are trying to model. In this case, we will consider a grid, defined in any system of coordinates, for which the computational representation does not change. Rectangular (Cartesian) or polar coordinates are logically rectangular, since their computational representations are the same. In this case, we will only handle a 1-D representation, and will assume a constant separation between the values arising from the discretization of the quantity of interest (in this case, space); i.e., we will assume it is a uniform grid. Finally, we will refer to the values arising from the discretization as **nodes**. Nodal grids are different from staggered grids, as discussed in Chapter 4.

We will introduce a succession of code fragments (or snippets) that can be thought of as stages in the process of building a complete code. We will explain the importance of each individual snippet of code in a way that readers with almost no experience with programming can have a sense of how to progressively design and implement an algorithmic solution to a given problem of interest.

We begin by defining the class data fields or **attributes** (sometimes called **data members** or **members variables**). Since we are discretizing a spatial interval $[a, b]$ in steps of Δx, it should be clear that those three values are of interest. Let us define a preliminary model of the class, as follows:

```
1   class LogicallyRectangular1DUniformNodalGrid {
2     double west_bndy_;
3     double east_bndy_;
4     double step_size_;
5   };
```

Notice the definition's syntax.

We are also interested in the physical data the grids handle by means of their nodes. According to the theory presented thus far, we discretize physical quantities of interest in these nodal positions (projected faces). However, we would like to create a class to represent these stored values. Let us define the class `Node1D`, as follows:

```
1   class Node1D {
2     public:
3       Node1D();
4     private:
```

```
5      double coordinate_;
6      double pressure_;
7   };
```

Therefore, we can extend our class modeling of a grid to include a collection of nodes, as follows:

```
1   class LogicallyRectangular1DUniformNodalGrid {
2       Node1D *nodes_;
3       double west_bndy_;
4       double east_bndy_;
5       double step_size_;
6   };
```

5.2.1 Access Specification and Implementation Hiding

At this point, we are still concerned about how to consider implementation hiding, since we stated earlier that the lack of it was a major concern in structured programming.

We will refer to the class operations as **member functions** (sometimes called **methods**). These are defined within the classes, thereby establishing an intimate relationship among them, which is defined as **data encapsulation**. Encapsulation is a prime feature of OOP. We will define two operations for the construction of these objects. In OOP, these creation operations are called **constructors**. Similarly, we will specify how to "destruct" an object of this class. Analogously, in OOP, these operations are called **destructors**. Their implementation can be done, as follows:

```
1   class LogicallyRectangular1DUniformNodalGrid {
2       public:
3           // Constructors:
4           LogicallyRectangular1DUniformNodalGrid();
5           LogicallyRectangular1DUniformNodalGrid(double aa, double bb,
                   double dx);
6           // Destructors:
7           ~LogicallyRectangular1DUniformNodalGrid();
8       private:
9           Node1D *nodes_;
10          double west_bndy_;
11          double east_bndy_;
12          double step_size_;
13  };
```

It is worth clarifying that, when it comes to the actual code writing, defining a class is done in the so-called **header files**. The purpose of this is to instruct the compiler about the structure of that class. The actual implementation of the class methods can be done within the header, but it is usually a good practice (when possible) to write such implementations in the **source files**. This implementation represents the actual algorithmics underlying the functioning of the class methods.

Notice our introduction of two keywords within the class definition, besides the three operations we were interested in defining. The terms `public` and `private` are called **access specifiers**, and they represent the mechanism through which OOP supports implementation hiding. Only members labeled "public" are allowed to be referenced outside the **scope** of the class, whereas any attempt to access private members will result in an error. This way, OOP ensures (if done correctly by the programmer) that access to the attributes is made only by dedicated operations, thus ensuring the validity and consistency of the object they modify. In this fashion, we define the collection of operations that allow for the interaction with the class as the class **interface**.

From here, it should easy to conclude that any given collection of written modules or classes that can be **reused** to program further classes or applications, and that hide their implementation details, can be defined as an **application programming interface (API)**.

Now, let us define both of the constructors. The first constructor, or the **default constructor**, can be defined, as follows:

```
1  // Default constructor:
2  LogicallyRectangular1DUniformNodalGrid ::
       LogicallyRectangular1DUniformNodalGrid () {
3    nodes_ = NULL;
4    west_bndy_ = 0.0;
5    east_bndy_ = 0.0;
6    step_size_ = 0.0;
7  }
```

It is worth clarifying that it is a good design and programming practice to define default constructors, since these are invoked by the compiler in the absence of explicit constructors in our code. That is, for specific algorithmic scenarios that may arise when programming, the compiler requires the existence of a default constructor, which usually yields a null instance of the class. For example, some of these scenarios are:

1. When the constructor for a given class constructor does not explicitly call the constructor of one of its object-valued attributes.

2. When an object (or an array of objects) is declared or dynamically allocated with no arguments.

It should be clear that the latter snippet of code initializes a default instance of a grid. Furthermore, the usage of the **scope operator** (::) should be pointed out. This operator allows for the usage of the attributes of the class in the body of the operation, as it is depicted in the previous snippet of code. This is a sample of the usefulness of the naming conventions in the attributes, since they can be easily differentiated from their related input values, such as in the following definition for the second constructor:

```
// Second constructor:
LogicallyRectangular1DUniformNodalGrid::
    LogicallyRectangular1DUniformNodalGrid(
        double aa,
        double bb,
        double dx) {

    nodes_ = NULL;
    west_bndy_ = aa;
    east_bndy_ = bb;
    step_size_ = dx;
}
```

The destructor of the class is commonly defined when classes allocate any memory resources. Destructors, in general, should handle the deallocations of these resources so that no **memory leaks** occur. A more sophisticated constructor could compute the number of nodes n as a function of a, b, and Δx, and allocate memory for the array of nodes. Assuming that, we would define our destructor, as follows:

```
// Destructor:
LogicallyRectangular1DUniformNodalGrid::~
    LogicallyRectangular1DUniformNodalGrid() {

    if (nodes_ != NULL)
        free [] nodes_;
}
```

It is clear that, if the nodes array was never initialized, to try to deallocate it will result in a run-time error. Therefore, our destructor validates this possibility and acts accordingly.

Notice that the the nodes array is the only attribute which requires an explicit deallocation in the collection of nodes, since this is the only attribute for which the constructor may have dynamically allocated memory. We do not worry about a, b, and Δx, since these are static variables.

5.2.2 Mutators, Accessors, and Client Code

Once we have specified how to construct and destruct instances of our class (our objects), we are interested in determining how to interact with them, since it is clear that classes are meant to be used by **client codes** to perform the tasks they were created for.

Mutators or **set functions** are member functions which have the responsibility of modifying class attributes.

Mutators should not be confused with constructors. Even though constructors also modify the object's attributes, constructor are meant for initializing an object into a specific state, possibly requiring memory allocation; whereas mutators are only used to modify these attributes, perhaps more than once.

Let us illustrate this by defining all of the mutators we could possibly require: First, it is clear that we need to be able to modify the defining values of our grid, e.g., the west boundary value and so forth. We declare the prototypes for these mutators within the definition of our class (thus preserving encapsulation), as follows:

```
1  class LogicallyRectangular1DUniformNodalGrid {
2    public:
3      // Constructors:
4      LogicallyRectangular1DUniformNodalGrid ();
5      LogicallyRectangular1DUniformNodalGrid (double aa, double bb,
            double dx);
6      // Destructors:
7      ~LogicallyRectangular1DUniformNodalGrid ();
8      // Mutators:
9      void set_west_bndy (double aa);
10     void set_east_bndy (double bb);
11     void set_step_size (double dx);
12     void set_node (int idx, double pp);
13   private:
14     Node1D *nodes_;
15     double west_bndy_;
16     double east_bndy_;
17     double step_size_;
18     int number_nodes_;
19 };
```

The definition of **set_west_bndy**, for instance, could be done, as follows:

```
1  LogicallyRectangular1DUniformNodalGrid :: set_west_bndy (double aa)
      {
2
3    if (aa >= 0.0)
4      west_bndy_ = aa;
5  }
```

Notice how the operation ensures the correct specification of the attributes. An important subtlety is that of adding a pressure value to the i-th node, where i is the indexing value. For this, we would need to extend the definition of the **Node1D** class with its own mutator, as follows:

```
1    class Node1D {
2      public:
3        // Default constructor:
4        void Node1D ();
5        // Mutators:
6        void set_pressure (double pp) {
```

```
7
8    if (pp >= 0.0)
9      pressure_ = pp;
10       }
11     private:
12         double coordinate_;
13         double pressure_;
14   };
```

By extending the `Node1D` class as in the latter snippet of code, we can then instruct the mutator in our grid class (which would act as the client code for the node class) to modify values in the i-th node, as follows:

```
1  LogicallyRectangular1DUniformNodalGrid :: set_node(int idx, double
       pp) {
2
3    if (0 < idx && idx < number_nodes_) {
4      if (pp > 0.0) {
5        nodes_[idx - 1].set_pressure(pp);
6      }
7    }
8  }
```

Note that i should also be less than the number of nodes n, which was already considered as an attribute of the grid class. Also notice the use of the **access operator** (.), which, analogously to structures in structured programming, permits access to the members, provided their access specification allows it. That is, if the member function **set_pressure** were private, line 5 in the previous snippet of code would yield an error.

Similarly, **accessors** or **get functions** are member functions which have the responsibility of access class attributes, in order for the clients to inquire about them. Accessors could be used from more complex member functions to provide insight about the structure of the object of interests. For example, they could be used from **printing functions**, in order to provide the client code with the capability of printing interesting information about the object Similarly, they could be used from **viewing functions**, which are functions that can access graphic APIs to help visualize the object of interest. This is useful when dealing with 3-D meshes or complicated graphs.

Now, let us show how we can create accessors we may need for the grid class. We might be interested in printing its structure, to validate whether the discretization is correct. We can do this from the client code, as follows:

```
1  LogicallyRectangular1DUniformNodalGrid *grid;
2  int ii;
3
4  cout << "[" << grid->west_bndy() << " ";
5  for (ii = 0; ii < grid->number_nodes(); ii++) {
6    cout << grid->nodal_value(ii) << " ";
7  }
```

```
8 cout << "[" << grid->east_bndy() << "]" << endl;
```

First, notice the introduction of the notation:

$$(*aa).pp() \equiv aa\text{->}pp().$$ (5.1)

The latter notational convention is used for accessing members through pointers to classes. Also, notice that the latter snippet of code shows that we would require four different accessors, which are of relatively simple implementation. For example, take the following implementation:

```
1 int LogicallyRectangular1DUniformNodalGrid :: number_nodes() {
2
3   return number_nodes_;
4 }
```

Once again, a subtlety arises when accessing the nodal values, since these are part of the node class. We would have to provide this class with the respective accessors so that the **client class**, in this case the grid class, can make appropriate use of it, in order to provide the required functionality.

Let us now present an example of how client codes can interact with the defined class in order to provide information on how to discretize the spatial distribution of the pressure values:

```
1  // Required variables:
2  double aa;
3  double bb;
4  double step_size;
5  LogicallyRectangular1DUniformNodalGrid *grid;
6  int ii;
7
8  // Initializations:
9  aa = 0.0;
10 bb = 1.0;
11 step_size = 0.2;
12
13 // Creation of the grid:
14 grid = new LogicallyRectangular1DUniformNodalGrid(aa, bb,
        step_size);
15
16 // Creation of the grid's values:
17 grid->assign_values(&pressure_distribution_function);
18
19 // Printing the grid:
20 cout << "[" << grid->west_bndy() << " ";
21 for (ii = 0; ii < grid->number_nodes(); ii++) {
22    cout << grid->nodal_value(ii) << " ";
23 }
24 cout << "[" << grid->east_bndy() << "]" << endl;
```

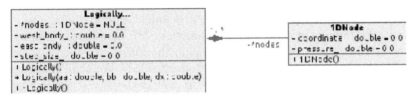

Figure 5.1: UML modeling of a 1-D grid and its nodes. Its details are explained in §5.3.

As is easily inferred, we can always add new functionalities to the collection of classes we use to write computer programs, or, again: our API; for as long as necessary. However, a correct modeling of the system of interest will yield a concise initial collection of requirements. In the next section (§5.3), we will explain the basics of modeling a system under an object-oriented approach. These modeling concepts represent a very important and fundamental aspect of API design.

5.3   Object-Oriented Modeling and UML

Object-orientation as a programming paradigm provides features that allow the computational modeling of any system of interest. In the interest of standardizing this modeling process, the **unified modeling language (UML)** provides semantic and semiotic mechanisms that allow developers to represent any entity of interest in a precise, and, more importantly, an intuitive manner.

For simplicity's sake, we will introduce one of the most important tools that UML provides for modeling in object-oriented design: the **class diagram**. Notice that though the UML provides more tools for complete studies in software engineering, however, these would be outside of the intended scope of this chapter. Some important examples of the omitted tools include the **activity diagram** and the **sequence diagram**.

Figure 5.1 shows the UML representation of the problem of modeling a 3-D grid, as we have been explaining throughout this chapter. The first aspect to notice is the representation of a class. In the UML, classes are represented as a rectangle with three sections. The top section contains the name of the class. Notice that it should be centered horizontally and boldfaced. The middle section contains the class attributes, and the bottom section contains the class operations (or methods).

When it comes to the attributes, translating the semantics of their specification, and, therefore, the semiotics of the figure, from the specification provided in C++ should be straightforward. Notice, however, that we have specified their visibility by using a + sign when it is public, and a − sign when it is private. An important observation is that the UML permits the suppression of the attributes and operations sections in the diagrams, when association among the classes is the only information intended. We shall make use of this valuable feature later in Chapter 6, to explain the interaction among the classes of a specific subset in our API. Such a diagram is said to be an **elided diagram**. When it comes to the operations, again, the diagram should be self-explanatory.

5.3.1   Relationships Among Classes

An important characteristic of considered classes is how they interact. The UML explains the mechanics of this interaction by means of the relationships it defines. These relationships can be classified as **instance-level relationships**, **class-level relationships**, and **general relationships**. We do not intend this section to be a thorough explanation of this. We are more interested in depicting the diverse nature of the interaction among classes.

Examples of instance-level relationships:

1. External links: Basic relationship among objects.

2. Association: Family of links.

3. Aggregation: A variant of the "has a" relationship. It is an association that represents a part-whole or part-of relationship.

4. Composition: A stronger aggregation. This represents the "is composed of" relationship. For example, the compound class is essential for the existence of the conformed class. Figure 5.1 shows an example of composition, since it is a design decision to specify that a grid is composed of nodes and that there is no such thing as a grid without nodes.

Examples of class-level relationships:

1. Generalization: This represents the "is a" relationship. We will explain it in detail in §5.4.

2. Realization: A relationship between two elements in which one element (the client) realizes (implements or executes) the behavior that the other element (the supplier) specifies. This should not be a strange concept, since we have already explained how client codes realize the functionalities provided by a class.

Finally, examples of general relationships:

1. Dependency: A weaker form of relationship which indicates that one class depends on another because it uses it at some point of time.

5.3.2   Multiplicity of a Relationship

When defining any relationship, an important aspect is the number of instances participating. This is called the **multiplicity** of the relationship. Figure 5.1 shows that, in a relationship between a grid and its nodes, any given grid can have one or more nodes; that is, we do not accept a grid with zero nodes. Notice again that this is nothing but a design decision. Examples of applications in which this is not necessary true are abundant.

The following table summarizes the possible multiplicities to be considered when modeling in the UML:

Table 5.1: Summary of possible multiplicities when modeling in the UML.

Symbol	Meaning
0	None
1	One
m	An integer value
0..1	Zero or one
m,n	m or n
$m..n$	At least m, but no more than n
*	Any nonnegative integer (zero or more)
0..*	Zero or more (identical to *)
1..*	One or more

Finally, an important concept will be introduced, and is shown in Figure 5.1. The word *nodes_ is a **role name**; role names identify the role that, in this case, the class Node1D plays within the definition of the grid class. (It should be clear that we decided not to write the entire name of the class in the diagram, simply for layout purposes.)

5.4   Inheritance and Polymorphism

To end this introductory chapter, we present two important and related concepts in OOP. Namely, those that allow us to "program in the general" rather that to "program in the specific."

In §5.1 we presented an example which included the necessity of handling 2-D points in certain applications. We mentioned that, in this case, structured programming would require the creation of a new structure, i.e., 2DPoint, to achieve this purpose, thus forcing the programmer to practically rewrite the code for the 3-D case (since it can be easily seen that the 2-D case is a specific instance of the 3-D case.)

Object-oriented programming presents **inheritance** as a mechanism to enforce code reutilization. Inheritance is the relationship that exist between a class with certain properties and its **base class**. In this case, the **derived class** enhances the base class by particularizing, thus extending its capabilities.

In Figure 5.2 we see that, if we were interested in supporting one, two, and tridimensional nodes, it would be more efficient to create an **interface class** to specify the general structure of a given node; that is, an indexation integer, for example, whereas the derived classes would contain the specification of the coordinates required in each dimensional scenario.

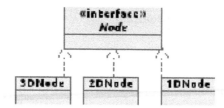

Figure 5.2: The UML modeling of an inheritance relationship.

Notice that, from the latter explanation, the concept of an interface class can be established to be any class whose purpose is solely to pose as a base class, and not to be instantiated. This contributes to software reutilization, which results in the boosting of developmental efficiency and overall accuracy of the modeling process.

5.4.1   Polymorphism and Operator Overloading

Our final important concept in OOP is that of **polymorphism**. Polymorphism is the ability to create algorithmic entities that have more than one form. There are several theories and practices of polymorphism within OOP. Some examples include

1. Operator overloading

2. Functions overloading and templates

3. Inheritance

4. Virtual functions

We are not interested in providing a thorough discussion of these concepts. However, we will briefly discuss the most important aspects of **operator**

overloading. Operator overloading is the capability of providing different implementations to operators, depending on their arguments.

For example, reconsider our `3DPoint` ADT, previously introduced in §5.1. We can easily extent this ADT to the following class:

```
1  class Point3D {
2    public:
3      // Operations:
4      Point3D operator=(Point3D aa) {
5        Point3D tmp = *this;
6
7        tmp.set_xx_coordinate(aa.xx_coordinate());
8        tmp.set_yy_coordinate(aa.yy_coordinate());
9        tmp.set_zz_coordinate(aa.zz_coordinate());
10       tmp.set_norm2(aa.norm2());
11       return tmp;
12     }
13
14     // Default constructor:
15     Point3D() {
16       xx_coordinate_ = 1.0;
17       yy_coordinate_ = 1.0;
18       zz_coordinate_ = 1.0;
19       norm2_ = sqrt(1.0*1.0 + 1.0*1.0 + 1.0*1.0);
20     }
21
22     // Mutators
23     void set_xx_coordinate(double xx) {
24       xx_coordinate_ = xx;
25     }
26     void set_yy_coordinate(double yy) {
27       yy_coordinate_ = yy;
28     }
29     void set_zz_coordinate(double zz) {
30       zz_coordinate_ = zz;
31     }
32     void set_norm2(double nn) {
33       norm2_ = nn;
34     }
35
36     // Accessors:
37     double xx_coordinate(void) {return xx_coordinate_;}
38     double yy_coordinate(void) {return yy_coordinate_;}
39     double zz_coordinate(void) {return zz_coordinate_;}
40     double norm2(void) {return norm2_;}
41
42   private:
43     double xx_coordinate_;
44     double yy_coordinate_;
45     double zz_coordinate_;
46     double norm2_;
47 };
```

Now, suppose we were interested in adding a given pair of normalized 3-D points, as is common in some algorithms within computer graphics. We could create a couple of overloaded operations for this. Specifically, we could overload the **+** operator supporting the addition of a couple of 3-D points. Furthermore, we could overload the **˜** operator to normalize a given 3-D point. Finally, we could overload the operator **%** to normalize both of the points and then sum them. The following snippet of code creates the first operator:

```
1  Point3D operator+(Point3D aa, Point3D bb) {
2    Point3D *tmp;
3
4    tmp = new Point3D();
5    tmp->set_xx_coordinate(aa.xx_coordinate() + bb.xx_coordinate())
         ;
6    tmp->set_yy_coordinate(aa.yy_coordinate() + bb.yy_coordinate())
         ;
7    tmp->set_zz_coordinate(aa.zz_coordinate() + bb.zz_coordinate())
         ;
8    tmp->set_norm2(sqrt(tmp->xx_coordinate()*tmp->xx_coordinate() +
9         tmp->yy_coordinate()*tmp->yy_coordinate() +
10        tmp->zz_coordinate()*tmp->zz_coordinate()));
11   return *tmp;
12 }
```

Notice that the operators should also guarantee the consistency of the resulting object. For example, in the previously defined operation, we made sure that even though the sum addition operation is not meant for computing the norm of the point, since this is an attribute of the class, it is computed as a result of the new point being the sum of the existing points. Furthermore, we could overload another operator, say the **&** operator, to compute the norm of an existing 3-D point:

```
1  double operator&(Point3D aa) {
2    double tmp;
3
4    tmp = sqrt(_aa.xx_coordinate()*aa.xx_coordinate() +
5      aa.yy_coordinate()*aa.yy_coordinate() +
6      aa.zz_coordinate()*aa.zz_coordinate());
7    return tmp;
8  }
```

Based on the latter snippet of code, the normalization operator can be implemented, as follows:

```
1  Point3D operator˜(Point3D aa) {
2    Point3D *tmp;
3    double aux;
4
5    // Compute the norm of the given point:
6    aux = &(aa);
```

```
7    // Define each of the new components by the norm of the given
         point:
8    tmp = new Point3D();
9    tmp->set_xx_coordinate(aa.xx_coordinate()/aux);
10   tmp->set_yy_coordinate(aa.yy_coordinate()/aux);
11   tmp->set_zz_coordinate(aa.zz_coordinate()/aux);
12   // Compute the norm of the new point, thus ensuring its
         consistency:
13   tmp->set_norm2(&(*tmp));
14   return *tmp;
15 }
```

And finally, our normalized sum operator follows:

```
1 Point3D operator%(Point3D aa, Point3D bb) {
2    Point3D tmp;
3
4    tmp = ~aa + ~bb;
5    return ~tmp;
6 }
```

Is there any important assumption made in the previous code? There certainly is. Notice that the previous snippet of code, the operator = was assumed to be overloaded in order to be able to assign an instance of one 3-D point into another. This was actually done in the definition of the class; recall:

```
1  class Point3D {
2    public:
3      // Operations:
4      Point3D operator=(Point3D aa) {
5        Point3D tmp = *this;
6
7        tmp.set_xx_coordinate(aa.xx_coordinate());
8        tmp.set_yy_coordinate(aa.yy_coordinate());
9        tmp.set_zz_coordinate(aa.zz_coordinate());
10       tmp.set_norm2(aa.norm2());
11       return tmp;
12     }
13
14     (...)
```

An example of a client code implementing this operation, should be straightforward:

```
1 int main (void) {
2
3    Point3D aa;
4    Point3D bb;
5
6    cout << "Test of Point3D overloading operators." << endl;
7    cout << "a = " << aa << endl;
```

```
8    cout << "b = " << bb << endl;
9    cout << "a % b = " << aa % bb << endl;
10
11   return EXIT_SUCCESS;
12 }
```

The latter snippet yields the following output:

```
Test of Point3D overloading operators.
a = (1, 1, 1) with ||.|| = 1.73205

b = (1, 1, 1) with ||.|| = 1.73205

a % b = (0.57735, 0.57735, 0.57735) with ||.|| = 1
```

Once again, we have assumed the overload of the `<<` operator, in order to permit printing the result stored in **ans**. Overloading this operator can be done analogously, as in the previous examples:

```
1  ostream &operator<<(ostream &out, Point3D aa) {
2
3    out << "(" << aa.xx_coordinate() << ", ";
4    out << aa.yy_coordinate() << ", ";
5    out << aa.zz_coordinate() << ") with ||.|| = " << aa.norm2() <<
          endl;
6    return out;
7  }
```

Algorithm 1 summarizes the entire study of operator overloading. We present this so the reader can compile it and execute it:

Algorithm 1 Complete source file depicting the application of operator overloading in manipulating points in a 3-D space.

```
1  #include <cstdlib>
2  #include <cmath>
3  #include <iostream>
4
5  using namespace std;
6
7  class Point3D {
8    public:
9      // Operations:
10     Point3D operator=(Point3D aa) {
11       Point3D tmp = *this;
```

```
12
13        tmp.set_xx_coordinate(aa.xx_coordinate());
14        tmp.set_yy_coordinate(aa.yy_coordinate());
15        tmp.set_zz_coordinate(aa.zz_coordinate());
16        tmp.set_norm2(aa.norm2());
17        return tmp;
18     }
19
20     // Default constructor:
21     Point3D() {
22        xx_coordinate_ = 1.0;
23        yy_coordinate_ = 1.0;
24        zz_coordinate_ = 1.0;
25        norm2_ = sqrt(1.0*1.0 + 1.0*1.0 + 1.0*1.0);
26     }
27
28     // Mutators
29     void set_xx_coordinate(double xx) {
30        xx_coordinate_ = xx;
31     }
32     void set_yy_coordinate(double yy) {
33        yy_coordinate_ = yy;
34     }
35     void set_zz_coordinate(double zz) {
36        zz_coordinate_ = zz;
37     }
38     void set_norm2(double nn) {
39        norm2_ = nn;
40     }
41
42     // Accessors:
43     double xx_coordinate(void) {return xx_coordinate_;}
44     double yy_coordinate(void) {return yy_coordinate_;}
45     double zz_coordinate(void) {return zz_coordinate_;}
46     double norm2(void) {return norm2_;}
47
48  private:
49     double xx_coordinate_;
50     double yy_coordinate_;
51     double zz_coordinate_;
52     double norm2_;
53 };
54
55 // Sum of two points:
56 Point3D operator+(Point3D aa, Point3D bb) {
57   Point3D *tmp;
58
59   tmp = new Point3D();
60   tmp->set_xx_coordinate(aa.xx_coordinate() + bb.xx_coordinate())
       ;
61   tmp->set_yy_coordinate(aa.yy_coordinate() + bb.yy_coordinate())
       ;
62   tmp->set_zz_coordinate(aa.zz_coordinate() + bb.zz_coordinate())
       ;
63   tmp->set_norm2(sqrt(tmp->xx_coordinate()*tmp->xx_coordinate() +
64           tmp->yy_coordinate()*tmp->yy_coordinate() +
```

```
65          tmp->zz_coordinate()*tmp->zz_coordinate()));
66    return *tmp;
67  }
68
69  // Norm of a point:
70  double operator&(Point3D aa) {
71    double tmp;
72
73    tmp = sqrt( aa.xx_coordinate()*aa.xx_coordinate() +
74      aa.yy_coordinate()*aa.yy_coordinate() +
75      aa.zz_coordinate()*aa.zz_coordinate());
76    return tmp;
77  }
78
79  // Normalize a point:
80  Point3D operator~(Point3D aa) {
81    Point3D *tmp;
82    double aux;
83
84    // Compute the norm of the given point:
85    aux = &(aa);
86    // Define each of the new components by the norm of the given
87         point:
88    tmp = new Point3D();
88    tmp->set_xx_coordinate(aa.xx_coordinate()/aux);
89    tmp->set_yy_coordinate(aa.yy_coordinate()/aux);
90    tmp->set_zz_coordinate(aa.zz_coordinate()/aux);
91    // Compute the norm of the new point, thus ensuring its
             consistency:
92    tmp->set_norm2(&(*tmp));
93    return *tmp;
94  }
95
96  // Sum of two normalized:
97  Point3D operator%(Point3D aa, Point3D bb) {
98    Point3D tmp;
99
100    tmp = ~aa + ~bb;
101    return ~tmp;
102  }
103
104  ostream &operator<<(ostream &out, Point3D aa) {
105
106    out << "(" << aa.xx_coordinate() << ", ";
107    out << aa.yy_coordinate() << ", ";
108    out << aa.zz_coordinate() << ") with ||.|| = " << aa.norm2() <<
             endl;
109    return out;
110  }
111
112  int main (void) {
113
114    cout << "Test of Point3D overloading operators." << endl;
115    Point3D aa;
116    Point3D bb;
117
```

```
118    Grid  gg;
119
120    cout << "a = " << aa << endl;
121    cout << "b = " << bb << endl;
122    cout << "a % b = " << aa % bb << endl;
123
124    return EXIT_SUCCESS;
125 }
```

The reader can create a source file with the entire code, say `testpoints.cc`, and compile it by instructing the following on a terminal (located in the same working directory as the file):

```
g++ testpoints.cc -o testpoints
```

5.5   Concluding Remarks

In this chapter, we introduced the most important concepts in object-oriented programming. We presented how this programming paradigm evolved naturally from that of structured programming, as well as important concepts related to classes and their attributes and methods. Important terms, such as encapsulation and implementation hiding, which pose the fundamentals of APIs design were also considered. Our objective was to prepare the reader for the concepts addressed when explaining MTK, our own API implementing mimetic discretization methods, in the following chapter.

We also explained the basics of the UML, a standard for modeling systems of interest under an object-oriented approach. We illustrated all of these concepts by means of simplified but representative source code, written in both C and C++.

Finally, we explained important features of C++ as an object-oriented language; namely, inheritance and polymorphism. We also explained a particular type of polymorphism called operator overload.

5.6   Sample Problems

1. Implement a structured program, using C as the programming language, and which implements a `Point3D` ADT.

   (a) Specify and justify the attributes and the operations.

(b) Write a client code that creates and orders a finite collection of randomly specified 3-D points based on their norm.

(c) Normalize them all and print all of their norms to visually corroborate that the normalization has been successful.

2. Extend the previous problem to C++.

   (a) Specify and justify the classes, their attributes, and the operations.

   (b) Write a client code that performs the same operations as in Problem 1.

   (c) Justify your design by means of a UML class diagram.

3. Create a UML class diagram that models the following scenario:

   > A given mesh can be considered to be a collection of grids. Meshes can be one, two, or tridimensional in geometry. Each grid in every dimensional scenario can be thought as a collection of nodes. However, 2-D meshes should posses their collection of edges and 3-D meshes should also posses their collection of faces. In general, meshes are a collection of cells. Faces can be conceptualized as a collection of edges; edges as a collection of nodes. Grids can be nodal or staggered, and nodes can be boundary nodes or inner nodes; however, in higher-dimensional scenarios a node can be a boundary node, a corner node, and a distinction should be made between simple nodes and hybrid nodes; the latter being defined as in Chapter 4.

   The description is intended to be ambiguous up to a certain point, yielding open answers for this problem (similar to modeling problems in the real world). However, you should justify every design decision; stating its purpose and possible implications.

4. Write a C++ client code that uses the collection of classes defined in Problem 2. Think of an effective way of printing/visualizing 3-D, 2-D, and 1-D meshes.

5. Extend the `1DNodalGrid` class to support the staggering operator !, so that if `nodal_gg` is an instance of a nodal 1-D grid, for example, then `(!nodal_gg)->Print()` prints the staggered grid spawned by `nodal_gg`.

Chapter 6

Mimetic Methods Toolkit (MTK)

Computational science and interdisciplinary studies have emerged as cutting-edge fields in recent years; due, in large part, to the fact that their tools and concepts are particularly efficient at helping researchers analyze and define the extraordinarily diverse range of physical phenomena that comprise our complex natural environment. These physical phenomena are usually modeled as a set of ordinary, or partial differential (or integral) equations, and correspond to a particular conservation law. Therefore, the numerical methods used to solve and study these equations are of vital importance in the current paradigm of interdisciplinary science.

Several noteworthy projects are currently underway whose primary goal is the creation of a variety of numerical application programming interfaces (APIs) which will allow users to implement numerical schemes products of theoretical development in an intuitive way. APIs provide abstractions for the problems under consideration, as well as specify how clients should interact with the software components that implement the problem's solution [151]. API development is ubiquitous in modern software development, and its purpose is to provide a logical interface for the functionality of a component, while hiding the details of its implementation.

Note that, in this chapter, we utilize `true type` text when referring to any computer-related context.

If we consider a particular programming paradigm as implemented through a specific programming language, as, for example C++, an API will generally include the following elements [151]:

- **Headers**: A collection of `.h` header files that define the interface and allow client code to be compiled against that interface. Open source APIs also include the source code (`.cpp` or `.cc` files) for the API implementation.

- **Libraries**: One or more static or dynamic library files that provide an implementation for the API. Clients can link their code against these library files in order to add any required functionality the API will provide to their applications.

- **Documentation**: Overview information that describes how to use the API, often including automatically generated documentation for all classes and functions in the API.

Well-known examples of this are given in [152], [153], [154], [155], and [156]; the last being an important step toward the understanding of portable computational performance and empirical tuning. Furthermore, as computational frameworks are developed to explore new boundaries in high-performance computing, the accompanying effort of developing these libraries goes on, as exemplified in [157]. Specifically, important theoretical work has been done in the field of numerical solution of partial and/or ordinary differential equations, ranging from the study of data structures [115], [116], and the development of algorithms [132], to the construction of APIs assisting in the implementation of numerical schemes to write scientific applications [114].

The importance of a well-designed API is vital in computational science, since a balance has to be attained between achieving computational performance and intuitively educating the user; not only in utilizing the API, but also in the theoretical aspects that underly its design. In the field of mimetic discretization methods, such a balance is particularly necessary, since no library has been yet developed to assist in their use when constructing scientific and industrial applications for which the simulation of some physical phenomena is of interest.

For the purpose of filling this void, we have developed the Mimetic Methods Toolkit (MTK). The MTK is an API that allows the intuitive implementation of mimetic discretization methods for the resolution of partial differential equations, yielding numerical solutions that guarantee uniform order of accuracy all along the modeled physical domain while ensuring the satisfaction of conservation laws, thereby remaining faithful to the underlying physics of the problem. The library is fully developed in C++, thus exploiting all the well-known advantages of both object-oriented application models, as well as the extensive collection of data structure capabilities of this language. Another example of an API fully developed in C++ is [114]; [158], on the other hand, describes how to write efficient and portable serial or parallel C++ programs for solving partial differential equations on a single curvilinear grid or on a collection of curvilinear grids that form an overlapping grid.

6.1   MTK Usage Philosophy

"Simplicity" is the omnipresent philosophy underlying the design of the MTK. With this in mind, we consider Algorithm 2, which depicts the creation of two grids for comparing the attained solution of a particular problem. Note that, in order to highlight the most important aspects of usage and functionality for the algorithm under consideration, we have ignored some portions of code that are related to the mechanics of the language.

In the search for simplicity and usability, the MTK interfaces with [159] to

---

Algorithm 2 Defining 1-D uniform nodal grids in the MTK.

```
1    MTK_Number aa;
2    MTK_Number bb;
3
4    MTK_LR1DUniformNodalGrid *nodal_grid;
5    MTK_LR1DUniformKnownSolutionHolderGrid *solution_grid;
6    MTK_1DPlotter *plotter;
7
8    cout << "Creating a nodal grid and a solution holder grid." <<
         endl;
9
10   aa = 0.0;
11   bb = 5.0;
12
13   nodal_grid =
14     new MTK_LR1DUniformNodalGrid(aa, bb, time_step);
15   cout << "Nodal grid:" << endl;
16
17   nodal_grid->Print1DArray();
18
19   solution_grid =
20     new MTK_LR1DUniformKnownSolutionHolderGrid(aa, bb, time_step,
            &known_sol_function);
21   cout << "Solution holder grid:" << endl;
22
23   solution_grid->Print1DArray();
```

---

present visual results tailored to each users' needs, as exemplified in Algorithm 3, with Figure 6.1 depicting the attained plot.

Algorithm 3 Use of an additional provided method to solve a sample problem while customizing a visual output with the MTK.

```
1    aa = 0.0;
2    bb = 5.0;
3    time_step = 0.1;
4    initial_value = 1.0;
5
6    nodal_grid =
7      new MTK_LR1DUniformNodalGrid(aa, bb, time_step);
8    cout << "Nodal grid:" << endl;
9    nodal_grid->Print1DArray();
10
11   solution_grid =
12     new MTK_LR1DUniformKnownSolutionHolderGrid(aa, bb, time_step)
       ;
13   cout << "Solution holder grid:" << endl;
14
15   solution_grid->Print1DArray();
16
17   converge = solution_grid->SolveInitialValue(MTK_HEUN,
       initial_value, &rhs);
18
19   plotter = new MTK_1DPlotter(solution_grid->independent(),
20                               solution_grid->dependent(),
21                               solution_grid->number_of_nodes
                                 ());
22   save_plot = false;
23   props = new MTK_1DPlotProperties("t", "f(t)", "Attained
       solution", "linespoints", "red");
24   plotter->set_plot_properties(props, save_plot, MTK_PNG);
25   plotter->See();
```

As previously mentioned, the MTK was developed using an object-oriented applications model, which is particularly conducive to educating the user in the field of mimetic discretization methods. Figure 6.2 shows a simplified UML diagram, which **ostensibly shows the relationship between the classes which handle meshes and grids within the MTK.** Clearly, Figure 6.2 is highly educational, since it reveals the theoretical implications of the interaction between the conceptualized entities that represent an important component of the theory of mimetic discretization methods, as presented in this text.

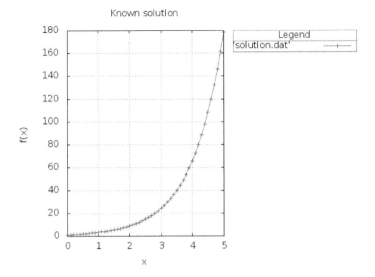

Figure 6.1: A visual result obtained from utilization of the MTK to solve for a sample problem under consideration.

6.2 Study of a Diffusive-Reactive Process Using the MTK

In this section, we introduce the example presented in §8.2 as a means of depicting how the user can utilize the MTK to solve for problems arising in CCUS. Recall that §8.2 introduces a steady 1-D diffusive-reactive process modeled as a dimensionless Poisson's equation, as described in (8.59). In order to be as concise as possible, we will describe the model, while simultaneously presenting the proposed algorithmic approach for solving it.

First, note the necessity of including the `mtk.h` header file in order to utilize the library. Similarly, related standard C++ headers are also considered:

```
1 #include <iostream>
2 #include <cstdlib>
3 #include <cmath>
4
5 #include "mtk.h"
6
7 using namespace std;
```

Now, define the functions implementing the source term of the system and the analytic solution (if known). Then, in cases like the example currently under consideration, we proceed, as follows:

```
1 MTK_Number source_term (MTK_Number xx) {
```

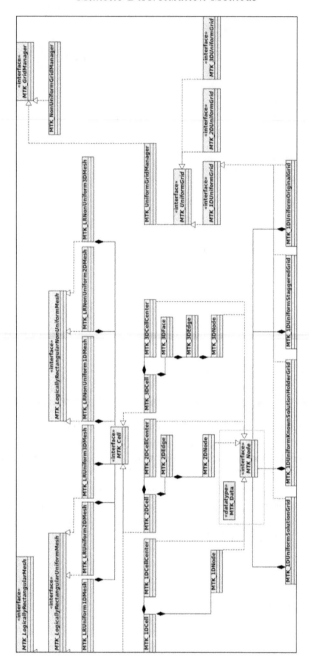

Figure 6.2: Simplified UML class diagram for the mesh and grid mechanisms within the MTK.

```
2
3   MTK_Number lambda;
4
5   lambda = -1.0;
6   return -(lambda*lambda)*exp(lambda*xx)/(exp(lambda) - 1.0);
7 }
8
9 MTK_Number known_solution (MTK_Number xx) {
10
11   MTK_Number lambda;
12
13   lambda = -1.0;
14   return (exp(lambda*xx) - 1.0)/(exp(lambda) - 1.0);
15 }
```

These should be compared with (8.58) and (8.59) in Chapter 8. For the example of interest, the following classes from the MTK should be used:

```
1    MTK_ToolManager *tools;
2    MTK_LR1DUniformStaggeredGrid *space;
3    MTK_LR1DUniformSourceGrid *source;
4    MTK_1DCGDirichletOperator *dir_comp;
5    MTK_1DCGNeumannOperator *neu_comp;
6    MTK_1DCGGradient *grad;
7    MTK_1DCGLaplacian *lap;
8    MTK_Steady1DProblem *problem_to_solve;
9    MTK_1DStencilMatrix *stencil_matrix;
10   MTK_LR1DUniformKnownSolutionHolderGrid *solution_grid;
11   MTK_1DPlotProperties *props;
12   MTK_1DPlotter *plotter;
```

The following auxiliary variables are subsequently considered in order to help define the problem of interest:

```
1    number_of_cell_centers = 50;
2    desired_order = 2;
3    lambda = -1.0;
4    west_bndy = 0.0;
5    east_bndy = 1.0;
6    alpha = -exp(lambda);
7    beta = (exp(lambda) - 1.0)/(lambda);
8    west_bndy_value = -1.0;
9    gamma = alpha;
10   delta = beta;
11   east_bndy_value = 0.0;
```

We are now able to create the required mimetic operators, as follows:

```
1 dir_comp = new MTK_1DCGDirichletOperator(desired_order,
2                                          number_of_cell_centers,
3                                          MTK_DENSE);
4
```

```
5  neu_comp = new MTK_1DCGNeumannOperator(desired_order,
6                                         number_of_cell_centers,
7                                         MTK_DENSE);
8
9  grad = new MTK_1DCGGradient(desired_order,
10                              number_of_cell_centers,
11                              MTK_DENSE);
12
13 lap = new MTK_1DCGLaplacian(desired_order,
14                             number_of_cell_centers,
15                             MTK_DENSE);
```

Similarly, we can define the problem of interest, as follows:

```
1  problem_to_solve = new MTK_Steady1DProblem(alpha, beta,
       west_bndy_value, gamma, delta, east_bndy_value);
```

We are now able to proceed with the creation of the required grids:

```
1  space = new MTK_LR1DUniformStaggeredGrid(west_bndy, east_bndy,
       number_of_cell_centers);
2  source = new MTK_LR1DUniformSourceGrid(space, west_bndy_value,
       east_bndy_value, source_term);
```

Note that the values of the source function (8.59) are provided, as well as the values for the discretized spatial coordinates. At this point, the MTK can now proceed with the creation of the stencil matrix and the system, which solution represents our desired solution. This is easily done by instructing:

```
1  stencil_matrix = new MTK_1DStencilMatrix(alpha, dir_comp,
2                                            beta, neu_comp,
3                                            grad, lap);
4  MTKSystem system(stencil_matrix, source);
5  system.Solve();
```

Note that the **Solve** member function implements a numerical method for solving the system. Since the operators were created using a dense representation for the storage format, a simple Gaussian elimination with backward substitution was utilized to solve the system [160]. However, full compatibility with known numerical linear algebra libraries is provided by the MTK. In fact, if we analyze the implementation of the constructor for the **MTK_1DStencilMatrix** class, we see that CBLAS [156] is used to perform all the required matrix operations, according to the definition of the stencil matrix M as a function of the mimetic matrix operators. The user can then visualize the required results graphically:

```
1  plotter = new MTK_1DPlotter(solution_grid->independent(),
2                              system.Solution(),
```

```
3                          solution_grid ->number_of_nodes () ) ;
4  save_plot = false ;
5  props = new MTK_1DPlotProperties ("x" ,
6                                    "u(x)" ,
7                                    "Computed  solution" ,
8                                    "points" ,
9                                    "green" ) ;
10 plotter ->set_plot_properties ( props , save_plot ,MTK_PNG) ;
11 plotter ->See () ;
```

Further, in order to visualize the known solution to this problem (for comparison purposes), we can now create a solution holder grid, such that:

```
1  solution_grid =
2  new  MTK_LR1DUniformKnownSolutionHolderGrid ( space , &known_solution
      ) ;
3  plotter =
4  new  MTK_1DPlotter ( solution_grid ->independent () , solution_grid ->
      dependent () , solution_grid ->number_of_nodes () ) ;
5  save_plot = false ;
6  props = new MTK_1DPlotProperties ("x" ,
7                                    "u(x)" ,
8                                    "Known  solution" ,
9                                    "lines" ,
10                                   "red" ) ;
11 plotter ->set_plot_properties ( props , save_plot ,MTK_PNG) ;
12 plotter ->See () ;
```

Finally, thanks to the tools within the MTK, we are able to numerically compare the error among the two solutions at hand:

```
1  tools = new MTK_ToolManager () ;
2  norm_diff_sol =
3  tools ->RelativeNorm2Difference ( solution_grid ->dependent () ,
4                                   system . Solution () ,
5                                   solution_grid ->number_of_nodes () ) ;
6  cout << "Relative  Norm  2  of  the  difference = " << norm_diff_sol
      << endl ;
7  return  EXIT_SUCCESS ;
```

Figure 6.3 shows the resulting plot of the known solution of the example problem. Similarly, Figures 6.4 and 6.5 depict the computed solution as the number of cells increase.

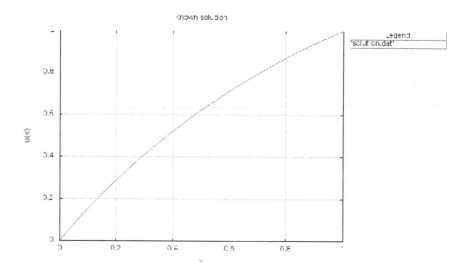

Figure 6.3: Known solution to the sample problem, visualized by means of the MTK visualization mechanisms.

Finally, Algorithm 4 depicts the complete source file as it should be written, so the reader can duplicate the results.

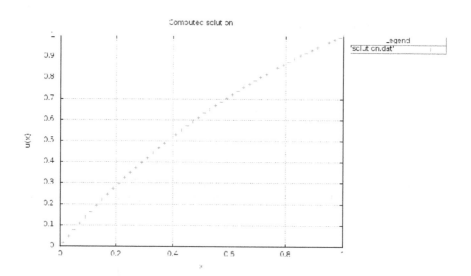

Figure 6.4: Computed solution of the sample problem using the MTK with 50 cells.

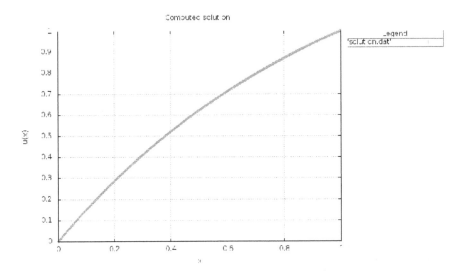

Figure 6.5: Computed solution of the sample problem using the MTK with 500 cells.

Algorithm 4 Complete source file depicting the resolution of a diffusive-reactive process using the MTK.

```
 1  #include <iostream>
 2  #include <cstdlib>
 3  #include <cmath>
 4
 5  #include "mtk.h"
 6
 7  using namespace std;
 8
 9  MTK_Number source_term (MTK_Number xx) {
10
11    MTK_Number lambda;
12
13    lambda = -1.0;
14    return -(lambda*lambda)*exp(lambda*xx)/(exp(lambda) - 1.0);
15  }
16
17  MTK_Number known_solution (MTK_Number xx) {
18
19    MTK_Number lambda;
20
21    lambda = -1.0;
22    return (exp(lambda*xx) - 1.0)/(exp(lambda) - 1.0);
23  }
24
25  int main () {
```

```
26
27    bool save_plot;
28
29    MTK_Number west_bndy;
30    MTK_Number east_bndy;
31    MTK_Number alpha;
32    MTK_Number beta;
33    MTK_Number west_bndy_value;
34    MTK_Number gamma;
35    MTK_Number delta;
36    MTK_Number east_bndy_value;
37    MTK_Number lambda;
38
39    MTK_Number norm_diff_sol;
40
41    int number_of_cell_centers;
42    int desired_order;
43
44    MTK_ToolManager *tools;
45    MTK_LR1DUniformStaggeredGrid *space;
46    MTK_LR1DUniformSourceGrid *source;
47    MTK1DCGDirichletOperator *dir_comp;
48    MTK_1DCGNeumannOperator *neu_comp;
49    MTK_1DCGGradient *grad;
50    MTK_1DCGLaplacian *lap;
51    MTK_Steady1DProblem *problem_to_solve;
52    MTK_1DStencilMatrix *stencil_matrix;
53    MTK_LR1DUniformKnownSolutionHolderGrid *solution_grid;
54    MTK_1DPlotProperties *props;
55    MTK_1DPlotter *plotter;
56
57    cout << "Example driver #1." << endl;
58    cout << "A Diffusion-Reaction (Poisson) Driver." << endl;
59    number_of_cell_centers = 60;
60    desired_order = 2;
61    lambda = -1.0;
62    west_bndy = 0.0;
63    east_bndy = 1.0;
64    alpha = -exp(lambda);
65    beta = (exp(lambda) - 1.0)/(lambda);
66    west_bndy_value = -1.0;
67    gamma = alpha;
68    delta = beta;
69    east_bndy_value = 0.0;
70
71    tools = new MTK_ToolManager();
72    dir_comp = new MTK_1DCGDirichletOperator(desired_order,
73                                             number_of_cell_centers,
74                                             MTK_DENSE);
75    dir_comp->Print(MTK_DENSE);
76    neu_comp = new MTK_1DCGNeumannOperator(desired_order,
77                                           number_of_cell_centers,
78                                           MTK_DENSE);
79    neu_comp->Print(MTK_DENSE);
80    grad = new MTK_1DCGGradient(desired_order,
81                                number_of_cell_centers,
```

```
82                                   MTK_DENSE) ;
83     grad->Print(MTK_DENSE) ;
84     lap = new MTK_1DCGLaplacian(desired_order ,
85                                 number_of_cell_centers ,
86                                 MTK_DENSE) ;
87     lap->Print(MTK_DENSE) ;
88
89     problem_to_solve = new MTK_Steady1DProblem(alpha , beta ,
           west_bndy_value , gamma, delta , east_bndy_value) ;
90     problem_to_solve->Print() ;
91
92     space = new MTK_LR1DUniformStaggeredGrid(west_bndy , east_bndy ,
           number_of_cell_centers) ;
93     cout << "Defined a LR 1D uniform staggered grid:" << endl << "\
           t" ;
94     space->Print1DArray() ;
95
96     stencil_matrix = new MTK_1DStencilMatrix(alpha , dir_comp ,
97                                              beta , neu_comp ,
98                                              grad , lap) ;
99     stencil_matrix->Print(MTK_DENSE) ;
100
101    source = new MTK_LR1DUniformSourceGrid(space , west_bndy_value ,
           east_bndy_value , source_term) ;
102    cout << "Defined a LR 1D uniform source grid:" << endl << "\t" ;
103    source->Print1DArray() ;
104
105    MTK_System system(stencil_matrix , source) ;
106    system.Print() ;
107    system.Solve() ;
108
109    solution_grid =
110    new MTKLR1DUniformKnownSolutionHolderGrid(space ,
111                                              &known_solution) ;
112    cout << "Solution holder grid:" << endl ;
113    solution_grid->Print1DArray() ;
114
115    plotter = new MTK_1DPlotter(solution_grid->independent() ,
116                                solution_grid->dependent() ,
117                                solution_grid->number_of_nodes()) ;
118    save_plot = false ;
119    props = new MTK_1DPlotProperties("x" , "u(x)" ,"Known solution" ,
120                   "lines" , "red") ;
121    plotter->set_plot_properties(props ,save_plot ,MTKPNG) ;
122    plotter->See() ;
123
124    plotter = new MTK_1DPlotter(solution_grid->independent() ,
125                                system.Solution() ,
126                                solution_grid->number_of_nodes()) ;
127    save_plot = false ;
128    props = new MTK_1DPlotProperties("x" ,
129                                     "u(x)" ,
130                                     "Computed solution" ,
131                                     "linespoints" ,
132                                     "blue") ;
133    plotter->set_plot_properties(props ,save_plot ,MTKPNG) ;
```

```
134   plotter ->See () ;
135
136   norm_diff_sol =
137   tools ->RelativeNorm2Difference ( solution_grid ->dependent () ,
138                         system . Solution () ,
139                         solution_grid ->number_of_nodes () ) ;
140   cout << "Relative Norm 2 of the difference = " << norm_diff_sol
         << endl ;
141   return EXIT_SUCCESS ;
142 }
```

6.3   Collaborative Development of the MTK: Flavors and Concerns

Another important aspect of our upcoming work is the collaborative development of the MTK. To that end, we propose two mechanisms for keeping control of the contributions to the API. The first mechanism is the **MTK's flavors**. Since the MTK's numerical core is written in C++, but the toolkit is intended to keep growing, diverse computational needs have to be taken into account; the result being the following "flavors" or APIs related to the MTK:

1. CMTK: C wrappers collection for the MTK; intended for sequential computations.

2. FMTK: Fortran 90 wrappers collection for the MTK; intended for sequential computations.

3. PyMTK: Python wrappers collection for the MTK; intended for sequential computations.

4. MMTK: MATLAB wrappers collection for the MTK; intended for sequential computations.

5. RMTK: R wrappers collection for the MTK; intended for sequential computations.

6. PMTK: Parallel extension for the MTK for distributed computing using MPI and OpenMP.

7. CuMTK: CUDA compatible extension for the MTK using MPI and CUDA.

The second mechanism is the conceptualization of the **MTK concerns**. Since collaborative development is key to achieving the level of generality desired for our library, we have divided the source code according to the designated purpose they possess within the API. These divisions (or "concerns") are hierarchically related by layers, and through the dependence of use among them; that is, one concern is said to depend on another if the classes it includes rely on the classes the latter includes. Note that this relation can be symmetrical, since two concerns may depend upon each other. Figure 6.6 depicts these concerns, as well as the interdependence between them.

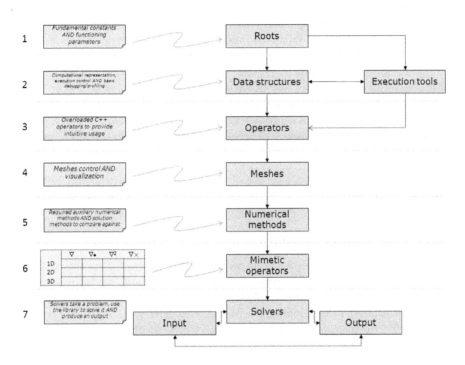

Figure 6.6: MTK concerns.

6.4   Downloading the MTK

To download the Mimetic Methods Toolkit (MTK), the reader can go online to

     `http://www.csrc.sdsu.edu/mimetic-book/`

This website contains all the information related to the project, including development news, information on the authors, and the API distribution.

To download, just access the website, and click on **Downloads**. The list of MTK flavors will appear, allowing the user to select the flavor of his/her preference. However, the flavors do require the core to work, which is listed first.

Just click on **Download** to get the `.tar.gz` file containing the API's core. Proceed similarly to obtain any of the other flavors.

From that point on, just follow the instructions provided in the `README` file.

This chapter presented a thorough description of one of the examples, which is intended to be understood by the reader, in order to grasp the MTK's usage philosophy.

6.5   Concluding Remarks

The purpose of this chapter is to provide a summarized perspective of the MTK's structure and philosophy of usage. To this end, we have supplied in depth analysis of several sample problems, and have explained the most important aspects of developing the MTK as an API.

Similarly, we have provided instruction on how to obtain an MTK distribution.

6.6   Sample Problems

1. Download and install the MTK. Perform the installation as instructed in the distributed `README` file. Are you able to execute example number five and interpret the attained results?

2. Modify example number five, so that you are able to time its execution for larger instances of the problem.

    (a) How long does it take to execute for different sizes of the problem?

(b) Are you able to write down the instance of the system of equations to be solved in this problem?

(c) Produce a plot of the attained solution.

3. How do the times improve if you specify to solve using ATLAS solvers instead of the default solvers used for Problem 2, for big cases?

4. Modify example five so that the right-hand side discretizing the forcing term is given by:

$$F(x) = \frac{2 \times 10^{-6}x}{\arctan(100)(1 + 1 \times 10^4 x^2)^2}. \tag{6.1}$$

For the west boundary, consider:

$$b_w = \frac{100}{\arctan(100)}, \tag{6.2}$$

and for the east boundary:

$$b_e = 1 + \frac{100}{\arctan(100)(1 + 1 \times 10^4)}. \tag{6.3}$$

If $\alpha = 1$ and $\beta = 1$ are coefficients of the Robin's boundary conditions, does the system of equations to be solved change?

5. Are you able to print the 1-D Castillo–Grone operators for different sizes?

Chapter 7

Nonuniform Structured Meshes

In this chapter we present a technique that uses mimetic operators to build mimetic schemes on nonuniform structured meshes, thereby achieving the same accuracy for the solution in the entire domain (as is the case for the Castillo–Grone operators).

By utilizing the idea proposed by D. Batista and J. Castillo [161], we are able to **locally** transform the various cells, as opposed to transforming the entire mesh all at once. The combined use of the 1-D DIV and GRAD high-order operators, and **local** transformations allows us to avoid the normal difficulty of keeping the order of accuracy p, for $p \geq 3$, which occurs when using a global reference mesh.

As in the case of the 1-D uniform staggered grid addressed earlier, we now consider a geometric mesh U, with the cells of this mesh being the intervals $[x_{i-1}, x_i]$, with $i = 1, 2, ..., n$. The edges of the cells in U are x_i, with $i = 0, 1, ..., n$, and are called **G-points**, because they are the points at which we define the gradient operator **G**. We also define the divergence operator **D** at the centers that can be described as: $x_{i+1/2}$ of these cells, calling them **D-points**, and $x_{i+1/2} = (x_i + x_{i+1})/2$.

As with the uniform grid, the complete set of all **G-points** and **D-points** is called a **staggered grid**. The cell $[x_{i-1}, x_i]$ is referred to as having the **cell number** $(i - 1/2)$. For the **second-order**, one-dimensional operators **D** and **G**, we must define the elements CD and CG (which are explained in the following section).

7.1 Divergence Operator

In order to calculate the divergence at $x_{i+1/2}$, with $i = 0, 1, ..., (n-1)$, we must first consider a map from $CD = [0, 1]$ to cell $(i + 1/2)$, which is given by:

$$x(\xi) = (x_{i+1} - x_i)\xi + x_i, \ 0 \leq \xi \leq 1. \tag{7.1}$$

Observe that $x(1/2) = (x_i + x_{i+1/2}) = x_{i+1/2}$, and that

$$J_x = \frac{dx}{d\xi}(\xi) = x_\xi = (x_{i+1} - x_i) = x(1) - x(0) \text{ for } 0 < \xi < 1, \tag{7.2}$$

In particular, we have

$$J_{x_{i+\frac{1}{2}}} = \frac{dx}{d\xi}(1/2) = x(1) - x(0). \qquad (7.3)$$

Also, by defining $\breve{v}(\xi) = v(x(\xi))$, we get $\breve{v}_\xi = v_x x_\xi$ and $v_x = \breve{v}_\xi / x_\xi$.

The introduction of the symbol J_x is motivated by the standard formula allowing us to evaluate the definite integrals over $[x_i, x_{i+1}]$ in terms of the definite integrals over $[0, 1]$:

$$\int_{x_i}^{x_{i+1}} \frac{dv}{dx} dx = \int_0^1 \frac{dv}{dx} J_x d\xi \approx \left(\frac{dv}{dx} J_x\right)(x_{i+1/2})[1 - 0] = \frac{dv}{dx}(x_{i+1/2}) J_{x_{i+1/2}}.$$

$$(7.4)$$

On the other hand, the fundamental theorem of calculus implies that

$$\int_{x_i}^{x_{i+1}} \frac{dv}{dx} dx = v(x_{i+1}) - v(x_i), \qquad (7.5)$$

so that

$$\frac{dv}{dx}(x_{i+1/2}) \approx (v(x_{i+1}) - v(x_i)) \frac{1}{J_{x+1/2}}. \qquad (7.6)$$

The 3-D version is used for the mapping of a 3-D cell onto the unit cube $CD_3 = [0, 1] \times [0, 1] \times [0, 1]$:

$$x(\xi) = (x_{i+1} - x_i)\xi + x_i, \ 0 \le \xi \le 1, \qquad (7.7)$$
$$y(\eta) = (y_{j+1} - y_j)\eta + y_j, \ 0 \le \eta \le 1, \qquad (7.8)$$
$$z(\zeta) = (z_{k+1} - z_k)\zeta + z_k, \ 0 \le \zeta \le 1. \qquad (7.9)$$

The fundamental theorem of calculus is now replaced by Gauss' divergence theorem, and by letting det J stand for the determinant of the Jacobian matrix for the abovementioned map, we get

$$\int_\Omega \nabla \cdot \mathbf{v} \, dx dy dz = \int_\sigma \mathbf{n} \cdot \mathbf{v} \, dv$$

$$= \int_{CD_3} (\nabla \cdot \mathbf{v}) \det J \, d\xi d\eta d\zeta$$

$$\approx \{(\nabla \cdot \mathbf{v}) \det J\}(x_{i+1/2}, y_{j+1/2}, z_{k+1/2})(1 - 0)^3.$$

which implies the following approximation for the divergence at the center of the 3-D cell

$$(\nabla \cdot \mathbf{v})(x_{i+1/2}, y_{j+1/2}, z_{k+1/2}) \approx \frac{\int_\sigma \mathbf{n} \cdot \mathbf{v} \, d\sigma}{\det J(x_{i+1/2}, y_{j+1/2}, z_{k+1/2})}. \qquad (7.10)$$

Our divergence operator, which is an $n \times (n+1)$ matrix acting upon a $(n+1) \times 1$ matrix $\mathbf{v}$ for nonuniform 1-D meshes looks like

$$\mathbf{Dv} = \begin{bmatrix} (\mathbf{Dv})(x_{1/2}) \\ (\mathbf{Dv})(x_{3/2}) \\ \vdots \\ (\mathbf{Dv})(x_{n-1/2}) \end{bmatrix}, \tag{7.11}$$

where $\mathbf{Dv}$ equals

$$\begin{bmatrix} -\frac{1}{J_{x_{1/2}}} & \frac{1}{J_{x_{1/2}}} & 0 & \cdots\cdots & & \cdots & 0 \\ 0 & -\frac{1}{J_{x_{3/2}}} & \frac{1}{J_{x_{3/2}}} & 0 & \cdots & & \cdots & 0 \\ \vdots & & \ddots & & \ddots & & & \vdots \\ 0 & \cdots & & \cdots & 0 & -\frac{1}{J_{x_{n-3/2}}} & \frac{1}{J_{x_{n-3/2}}} & 0 \\ 0 & \cdots & & \cdots\cdots & & 0 & -\frac{1}{J_{x_{n-1/2}}} & \frac{1}{J_{x_{n-1/2}}} \end{bmatrix} \begin{bmatrix} v_0 \\ v_1 \\ \vdots \\ \vdots \\ v_{n-2} \\ v_{n-1} \\ v_n \end{bmatrix}. \tag{7.12}$$

Clearly, our divergence operator $\mathbf{D}$ for nonuniform meshes is the product of $(J_D)^{-1}$ times the fixed part of the divergence operator for **uniform meshes** presented in [142], where

$$J_D = DIAG\left(J_{x_{1/2}}, J_{x_{3/2}}, ..., J_{x_{n-3/2}}, J_{x_{n-1/2}}\right), \tag{7.13}$$

and $\mathbf{D}$ equals

$$\begin{bmatrix} \frac{1}{J_{x_{1/2}}} & 0 & 0 & \cdots & & 0 \\ 0 & \frac{1}{J_{x_{3/2}}} & 0 & \cdots & & 0 \\ \vdots & & \ddots & & & \vdots \\ 0 & & \cdots & 0 & \frac{1}{J_{x_{n-3/2}}} & 0 \\ 0 & & \cdots & 0 & 0 & \frac{1}{J_{x_{n-1/2}}} \end{bmatrix} \begin{bmatrix} -1 & 1 & 0 & 0 & \cdots & 0 \\ 0 & -1 & 1 & 0 & \cdots & 0 \\ \vdots & & \ddots & & & \vdots \\ 0 & \cdots & 0 & -1 & 1 & 0 \\ 0 & \cdots & 0 & 0 & -1 & 1 \end{bmatrix}. \tag{7.14}$$

7.2 Gradient Operator

To calculate the gradient at an interior **G**-point x_i, with $i = 1, ..., (n-1)$, and unequal distances from x_i to $x_{i-1/2}$ and $x_{i+1/2}$, we need to use two cells, cell $(i - 1/2)$ and cell $(i + 1/2)$, and consider a map from $CG = [-1, 0] \cup [0, 1]$ to those cells, as given by

$$x(\xi) = \frac{[x_{i-1/2} - 2x_i + x_{i+1/2}]}{(1/2)}\xi^2 + [x_{i+1/2} - x_{i-1/2}]\xi + x_i \text{ for } -0.5 \le \xi \le 0.5. \tag{7.15}$$

Observe that

$$\frac{dx}{d\xi}(0) = [x_{i+1/2} - x_{i-1/2}] = J_{x_i}, \qquad (7.16)$$

and $x(0) = x_i$, $x(-0.5) = x_{i-1/2}$ and $x(0.5) = x_{i+1/2}$.

As before, the gradient at the interior point x_i is calculated as

$$(\mathbf{G}f)(x_i) = f_x(x_i) = \left.\frac{\check{f}_\xi}{x_\xi}\right|_{\xi=0} \qquad (7.17)$$

and

$$(\mathbf{G}f)(x_i) \approx (f(x_{i+1/2}) - f(x_{i-1/2}))\frac{1}{J_{x_i}}. \qquad (7.18)$$

The boundary points x_0 and x_n require special treatment as well as the full extension of cells $(i - 1/2)$ and $(i + 1/2)$.

Using the approach presented in [142], we obtain

$$(\mathbf{G}f)(x_0) = f_x(x_0) = \left.\frac{\check{f}_\xi}{x_\xi}\right|_{\xi=-1} = \frac{[-\frac{8}{3}\check{f}(-1) + 3\check{f}(-0.5) - \frac{1}{3}\check{f}(0.5)]}{[-\frac{8}{3}x(-1) + 3x(-0.5) - \frac{1}{3}x(0.5)]} \qquad (7.19)$$

$$= [-\frac{8}{3}f(x_0) + 3f(x_{1/2}) - \frac{1}{3}f(x_{3/2})]\left(\frac{1}{J_{x_0}}\right),$$

where $J_{x_0} = -\frac{8}{3}x_0 + 3x_{1/2} - \frac{1}{3}x_{3/2}$.

The gradient at the boundary point x_n is acquired by transforming the cells $(n - 3/2)$ and $(n - 1/2)$ into CG.

The elements CD and CG have been defined as uniform reference elements, and they are where the actual approximations are carried out, and, together, they form the **reference set of cells (RSC)**. The RSC is independent of the number of points of the mesh, though its elements depend on the order of accuracy desired for the DIV and GRAD operators. Of course, the RSC also depends upon the type of elements employed for discretizing the physical domain, as well as the dimension of physical problems being modeled.

However, having established these additional aspects, the RSC remains unchanged while constructing new mimetic operators.

As a result, our gradient operator for nonuniform 1-D meshes is defined as

$$\mathbf{G}f = \begin{bmatrix} (\mathbf{G}f)(x_0) \\ (\mathbf{G}f)(x_1) \\ \vdots \\ (\mathbf{G}f)(x_n) \end{bmatrix}, \qquad (7.20)$$

where $\mathbf{G}f$ equals

$$
\begin{bmatrix}
-\frac{8/3}{J_{x_0}} & \frac{3}{J_{x_0}} & -\frac{1/3}{J_{x_0}} & 0 & \cdots & & \cdots & 0 \\
0 & -\frac{1}{J_{x_1}} & \frac{1}{J_{x_1}} & 0 & \cdots & & \cdots & 0 \\
\vdots & & \ddots & & & \ddots & & \vdots \\
0 & \cdots & & \cdots & 0 & -\frac{1}{J_{x_{n-1}}} & \frac{1}{J_{x_{n-1}}} & 0 \\
0 & \cdots & & \cdots & 0 & \frac{1/3}{J_{x_n}} & -\frac{3}{J_{x_n}} & \frac{8/3}{J_{x_n}}
\end{bmatrix}
\begin{bmatrix}
f_0 \\
f_{1/2} \\
\vdots \\
\vdots \\
\vdots \\
f_{n-3/2} \\
f_{n-1/2} \\
f_n
\end{bmatrix}.
\tag{7.21}
$$

As with the **D-operator**, our **G-operator** can also be rewritten as a product of two matrices: one symbolized by $(J_G)^{-1}$, and the other being the fixed part of the gradient operator for uniform meshes, where $J_G = DIAG\left(J_{x_0}, J_{x_1}, ..., J_{x_{n-1}}, J_{x_n}\right)$ and $\mathbf{G}$ equals

$$
\begin{bmatrix}
\frac{1}{J_{x_0}} & 0 & 0 & \cdots & & 0 \\
0 & \frac{1}{J_{x_1}} & 0 & \cdots & & 0 \\
\vdots & & \ddots & & & \vdots \\
0 & & \cdots & 0 & \frac{1}{J_{x_{n-1}}} & 0 \\
0 & & \cdots & 0 & 0 & \frac{1}{J_{x_n}}
\end{bmatrix}
\begin{bmatrix}
-8/3 & 3 & -1/3 & 0 & \cdots & 0 \\
0 & -1 & 1 & 0 & \cdots & 0 \\
\vdots & & \ddots & & & \vdots \\
0 & \cdots & 0 & -1 & 1 & 0 \\
0 & \cdots & 0 & 1/3 & -3 & 8/3
\end{bmatrix}.
\tag{7.22}
$$

As in the case of uniform meshes, the use of P and Q results in the boundary operator **B**. Now, recalling that $\mathbf{G}_{nu} = (J_G)^{-1}G_u'$, $\hat{\mathbf{D}}_{nu} = (J_{\hat{D}})^{-1}\hat{D}_u'$, $P_{nu} = P_u' J_\mathbf{G}$ and $Q_{nu} = Q_u'(J_{\hat{\mathbf{D}}})$, we can compute

$$
\begin{aligned}
\mathbf{B}_{nu} &= Q_{nu}\hat{\mathbf{D}}_{nu} + (P_{nu}\mathbf{G}_{nu})^T \\
&= Q_u'(J_{\hat{\mathbf{D}}})(J_{\hat{\mathbf{D}}})^{-1}\hat{D}_u' + (P_u'(J_\mathbf{G})(J_\mathbf{G})^{-1}\mathbf{G}_u')^T \\
&= Q_u'\hat{\mathbf{D}}_u' + (P_u'\mathbf{G}_u')^T = \mathbf{B}_u.
\end{aligned}
\tag{7.23}
$$

That is, the second-order boundary operator for nonuniform meshes **is the same operator obtained for the uniform case**, and is simpler than that obtained in [147].

Finally, the fixed parts and global operators from the nonuniform case exhibit the following properties:

1. **D** and **G** are centro-skew-symmetric

2. **D** and **G** are banded matrices

Additionally, the global conservation law

$$
h<\hat{\mathbf{D}}v, f>_Q + h<\mathbf{G}f, v>_P = <\mathbf{B}v, f>
\tag{7.24}
$$

is satisfied on nonuniform meshes. The procedure can also be repeated for 4th-order **G** and **B**.

The combination of adapted meshes with mimetic schemes (resulting from the use of high-order P_{nu} and Q_{nu}) has proven to be a valuable option for solving difficult boundary-layer problems. An even simpler example of this type can be found in [161].

7.3 Concluding Remarks

In this chapter we introduced the fundamental concepts for constructing mimetic approximations over nonuniform meshes. New elements were introduced to the theory of mimetic discretization methods: the reference set of cells (RSC), and the use of local transformations, as well as the basic tools for constructing mimetic operators over nonuniform meshes.

7.4 Sample Problems

1. Write a code that uses the Castillo–Grone Laplacian defined in §3.5.1 as $\breve{\mathbf{L}} = \mathring{\mathbf{D}}\breve{\mathbf{G}}$. Solve Problem 1 from Chapter 4 by means of a nonuniform mesh.

   (a) For the 1-D uniform case, compare the solution by means of FD and a mimetic method using a nonuniform mesh.

   (b) For the 1-D uniform case, compare the solution by means of a uniform and a nonuniform case.

   (c) Create a computer plot of the error as a function of grid refinement for the nonuniform mesh case. Is it second order?

2. Verify that the proposed simple nullity tests in §3.6 become the discrete analog for the equality of mixed partial derivatives when second-order accurate Castillo–Grone operators $\breve{\mathbf{G}}$ and the resulting D_2 are used.

3. Modify the previous exercise so that the right-hand side discretizing the forcing term is given by

$$F(x) = \frac{2 \times 10^{-6}x}{\arctan(100)(1 + 1 \times 10^4 x^2)^2}. \quad (7.25)$$

Consider, for the west boundary

$$b_1 = \frac{100}{\arctan(100)}, \quad (7.26)$$

and, for the east boundary

$$b_1 = 1 + \frac{100}{\arctan(100)(1 + 1 \times 10^4)}. \qquad (7.27)$$

Consider $\alpha = 1$ and $\beta = 1$ for the coefficient of the Robin boundary conditions. How do the systems of equations to be solved change now? Is there any important property to observe among them? What about the effect of grid refinement?

4. Propose an algorithm for the implementation of the Castillo–Grone 2-D divergence and 2–D gradient in nonuniform meshes. Can you compute the Castillo–Grone Laplacian? Can you propose a problem in which these are of utility?

5. Propose an algorithm for the implementation of the Castillo–Grone 3-D divergence and 3-D gradient in nonuniform meshes. Can you compute the Castillo–Grone Laplacian? Can you propose a problem in which these are of utility?

Chapter 8

Case Studies

A large number of time-dependent problems in physics and engineering are governed by some type of wave equation, which can be in the form of either a scalar or a vector PDE. Typically, boundary conditions must be added to the other initial conditions as a means of ensuring the uniqueness of the solution. Other types of problems, typically of the diffusion kind, are governed by PDEs of parabolic types, also requiring an initial condition. Steady state phenomena are usually governed by equations of elliptic type.

Some of the traditional finite difference methods that are used in the solution of a system of PDEs present an important drawback, which is related to the low order of accuracy used in the numerical treatment of boundary conditions, with respect to the discretization of the interior fields (this is true even in regular geometries such as a half-space.)

The fourth- and sixth-order $\check{\mathbf{D}}$ and $\check{\mathbf{G}}$ operators first proposed by Castillo and Grone (see [142]), ensure that the discretization of the boundary conditions is compatible (regarding both the physics and the accuracy) with the interior computation by avoiding any outer-domain or ghost points.

In this chapter we present application problems in which mimetic discretization methods have been successfully applied. To that end, we will also present a highly diverse collection of background knowledge.

8.1 Porous Media Flow and Reservoir Simulation

The field of hydrocarbon extraction has evolved to the point that it now requires the utilization of sophisticated mathematical tools. For example: The fluid displacement processes that takes place in porous media are related to both chemical and physical phenomena, as well as to changing fluid interfaces. As a result, the numerical modeling of the various intricate phenomena has more than a practical interest: it also has the potential for further development in the field of computational science. A survey of the variety of problems arising in this area has been compiled by R. E. Ewing (see [162]).

8.1.1 Darcy's Law and the Pressure Equation

The application of Darcy's law to modeling flows in reservoir simulation leads to elliptic PDEs that are akin to Poisson's equation with Robin boundary conditions.

If $\mathscr{K}$ denotes the permeability tensor of the rock (see Appendix B), and **u** the Darcy velocity, then the relationship between the flow rate and the pressure gradient (not taking gravity's effects into consideration) is given by **Darcy's law**:

$$\mathbf{u} = -\frac{1}{\mu}(\mathscr{K} \cdot \nabla p), \tag{8.1}$$

where μ is the viscosity of the fluid.

The Darcy velocity is the total discharge rate of the fluid divided by the cross-sectional area to flow, which is not the velocity which the liquid traveling through pores is experiencing. The **Pore velocity v** equals the Darcy velocity **u** divided by the porosity ϕ, since only a fraction of the total formation volume is available for flow, and $0 \leq \phi \leq 1$.

The intrinsic permeability is a function of the porosity of the rock formation, and, according to Darcy, is proportional to the square of the average pore diameter. The rock pores must be interconnected, in order for the incoming fluid to have available pathways through the formation. Pore geometry and rock constitution may result in **anisotropic permeability**.

In Equation (8.1), the diagonal tensor components k_x, k_y, and k_z are the permeability constants for the x-, y-, and z-directions, respectively. In the case of an isotropic medium, $k_x = k_y = k_z = k$ and $k_{ij} = 0$ for $i \neq j$.

In the case of a single fluid phase, if q denotes the mass flow rate per unit volume and ρ is the fluid density, then the steady state equation reflecting the conservation of mass leads to

$$\nabla \cdot \left(\frac{1}{\mu}(\mathscr{K} \cdot \nabla p) + \frac{q}{\rho} \right) = 0. \tag{8.2}$$

This is usually referred to as **the pressure equation**, and represents an elliptic PDE for a single fluid phase.

8.1.2 Diphasic Flow and Permeability

Let us assume that there are two types of immiscible fluids flowing in a porous medium: one is wet (w) and the other is not (n). We can now denote the saturation of the wet and non-wet fluid phases as S_w and S_n, respectively. Since these indicate the fraction of volume occupied in the flow, we have $S_w + S_n = 1$. We also have Darcy velocities denoted as $\mathbf{u}_w$ and $\mathbf{u}_n$, which satisfy

$$\mathbf{u}_w = -\frac{k_{rw}}{\mu_w}\nabla p_w, \tag{8.3}$$

$$\mathbf{u}_n = -\frac{k_{rn}}{\mu_n}\nabla p_n. \tag{8.4}$$

Here, it has been assumed that the formation is isotropic, so that a scalar magnitude k or absolute permeability can be defined (and the tensor $\mathcal{K} = kI$ in this case equals I, the identity matrix), and it models the empirical permeability exhibited by the rock for the seepage of a single-phase reference fluid; namely, water. This physical scalar parameter might vary from point to point within the rock, but will not depend upon directions, and we will write $k(\mathbf{x})$ in such a case. The difference between p_n and p_w is called the **capillary pressure**, and is expressed as $p_c = p_n - p_w$.

Here, we have μ_w and μ_n as the viscosities of the fluid phases, and k_w and k_n as the "effective" permeability constants. The relative permeability constants k_{rw} and k_{rn} are defined as

$$k_{rw} = \frac{k_w}{k}, \tag{8.5}$$

and

$$k_{rn} = \frac{k_n}{k}. \tag{8.6}$$

After defining the **average pressure** p between the two phases, with a total mobility of

$$\lambda(S_w) = \lambda = \frac{k_{rn}(S_w)}{\mu_n} + \frac{k_{rw}(S_w)}{\mu_w}, \tag{8.7}$$

we can now derive the following elliptic PDE:

$$\nabla \cdot \mathbf{u} = -\nabla \cdot (k(\mathbf{x})\lambda \nabla p) = \frac{q_n}{\rho_n} + \frac{q_w}{\rho_w}, \tag{8.8}$$

where $\mathbf{u}$ is the total fluid velocity, and constant phase densities, constant porosity, and incompressibility have been assumed for both the fluids and the rocks.

8.1.3 Generalization and the Material Tensor

Generally, $k(\mathbf{x})\lambda$ must be substituted for either $\mathcal{K}(x)$ or a **material tensor**, depending upon (k_w/μ_w) and (k_n/μ_n), which are important properties of the fluid, and, in the 2-D case, is a two-by-two positive-definite matrix:

$$\mathcal{K} = \begin{bmatrix} k_{11}(x) & k_{12}(x) \\ k_{21}(x) & k_{22}(x) \end{bmatrix}. \tag{8.9}$$

For staggered 2-D uniform grids, the grid tensor coefficient $\mathscr{K}$ needs to be defined at the centers of the edges of the 2-D cells, and also, when it comes to the boundaries of our rectangular 2-D staggered grid, the computation of $\mathscr{K}$ should be performed at the centers of the 1-D cell centers, which are projected onto the boundary segments that are coincident (or parallel) to the coordinate axis.

Let $\Omega = [x_L, x_U] \times [y_L, y_U]$ be a 2-D space region, and define: $\Omega^o = (x_L, x_U) \times (y_L, y_U)$. Also, let $\partial\Omega = \sigma$ denote the boundary of Ω.

Renaming $p = f$ and $F = \frac{q_n}{\rho_n} + \frac{q_w}{\rho_w}$, the source term, we see that our elliptic PDE can be written as

$$div(\mathscr{K} \cdot grad \; f) = -F \text{ on } \Omega^o. \qquad (8.10)$$

In the case of Robin's boundary condition at the boundary surface σ:

$$\alpha f + \beta < \mathscr{K} \cdot grad \; f, \mathbf{n} >= b \text{ for all } (x, y) \in \partial\Omega, \qquad (8.11)$$

where α is the scalar value defining the Dirichlet term, β is the scalar value defining the Neumann term, b is the given right-hand side function at the boundary $\partial\Omega$, and, as usual, $\mathbf{n}(x, y)$ denotes the outwardly directed unit normal at $(x, y) \in \partial\Omega$.

8.1.4 Example Using a Mimetic Method

For simplicity's sake, we now briefly describe uniform second-order accurate mimetic discretization schemes on 1-D grids; where $\mathscr{K}$ is equal to the identity matrix, so that $\Omega = [0, 1]$ and $\partial\Omega = \{0, 1\}$.

In this case, we obtain Poisson's equation (which is a typical elliptic PDE) for an unknown function f, as well as for the given function F, so that

$$-\nabla^2 f(x) = F(x) \text{ for all } x \in \Omega. \qquad (8.12)$$

Now, let us consider the **Robin problem** for constant values of α and β. In this case, F is a given right-hand side function and b is **given** at the boundary surface $\sigma = \partial\Omega$, so that the Robin condition must be satisfied, as follows:

$$(\alpha f + \beta < grad \; f, \mathbf{n} >)(x) = b(x) \text{ for } x \in \sigma. \qquad (8.13)$$

Next, we illustrate the construction of **a mimetic numerical scheme** by utilizing a 1-D uniform staggered grid for the discretization of the physical domain $\Omega = [0, 1]$, $\partial\Omega = \{0, 1\}$, so that Equations (8.12) and (8.13) become

$$-\frac{d^2 f}{dx^2}(x) = F(x) \text{ for } 0 < x < 1$$
$$\alpha f(0) - \beta f'(0) = lb$$
$$\alpha f(1) + \beta f'(1) = rb.$$

Here we have denoted the left and right boundary values of b by lb and rb, respectively.

Now, we consider the standard 1-D uniform staggered grid

$$(x_0, x_{1/2}, x_1, x_{3/2}, ..., x_{n-1/2}, x_n), \tag{8.14}$$

where $x_0 = 0$ and $x_n = 1$ are boundary nodes, and

$$x_{1/2}, x_{3/2}, ..., x_{n-1/2} \tag{8.15}$$

are the cell center nodes. We introduce the following suggestive notations:

- $f = f_{cb} = (f(x_0), f(x_{1/2}), ..., f(x_{n-1/2}), f(x_n))^T \in \mathbb{R}^{(n+2)}$,

- $f_c = (f(x_{1/2}), f(x_{3/2}), ..., f(x_{n-1/2}))^T \in \mathbb{R}^n$,

- $b = (lb, F(x_{1/2}), F(x_{3/2}), ..., F(x_{n-1/2}), rb)^T \in \mathbb{R}^{(n+2)}$.

Defining the operators L and $A \in \mathbb{R}^{(n+2)\times(n+2)}$, as well as the boundary operators $B \in \mathbf{R}^{(n+2)\times(n+2)}$ and $\breve{B}$ as before, we are now able to write the discrete mimetic versions of (8.12) and (8.13) as **single matrix equations**. Now, reverting to the physically motivated notation $f = p$, we obtain

$$(\alpha A + \beta BG - L)p = b, \tag{8.16}$$

where:

$$A = \begin{bmatrix} 1 & 0 & 0 & ... & & ... & 0 \\ 0 & 0 & 0 & ... & & ... & 0 \\ 0 & 0 & 0 & ... & & ... & 0 \\ 0 & 0 & 0 & ... & & ... & 0 \\ \vdots & \vdots & \vdots & \ddots & \ddots & \vdots & \vdots & \vdots \\ 0 & ... & & & ... & 0 & 0 & 0 \\ 0 & ... & & & ... & 0 & 0 & 0 \\ 0 & ... & & & ... & 0 & 0 & 0 \\ 0 & ... & & & ... & 0 & 0 & 1 \end{bmatrix}, \tag{8.17}$$

$$G = \frac{1}{h} \begin{bmatrix} -2 & 2 & 0 & 0 & 0 & ... & ... & 0 \\ 0 & -1 & 1 & 0 & 0 & ... & ... & 0 \\ 0 & 0 & -1 & 1 & 0 & ... & ... & 0 \\ \vdots & \vdots & \vdots & \ddots & \ddots & \vdots & \vdots \\ 0 & ... & ... & 0 & -1 & 1 & 0 & 0 \\ 0 & ... & ... & 0 & 0 & -1 & 1 & 0 \\ 0 & ... & ... & 0 & 0 & 0 & -2 & 2 \end{bmatrix}, \tag{8.18}$$

and

$$L = \begin{bmatrix} 0 & \cdots & 0 \\ & DG & \\ 0 & \cdots & 0 \end{bmatrix}. \tag{8.19}$$

Observe that DG contains the factor $1/h^2$ as a discrete Laplacian should.

Next, note that, in the discrete version of the IBP, the introduction of the weight matrices P and Q forces the change from B to $\breve{B}$, with B and $\breve{B} \in \mathbb{R}^{(n+2)\times(n+1)}$, with

$$\breve{B} = h\{\hat{Q}\hat{\breve{D}} + (P\breve{G})^T\}. \tag{8.20}$$

Now, recall that D and G have also changed to $\breve{D}$ and $\breve{G}$, respectively.

Since $\breve{G} \in \mathbb{R}^{(n+1)\times(n+2)}$, then $\breve{B}\breve{G} \in \mathbb{R}^{(n+2)\times(n+2)}$ as $\breve{L}$ and A do, with $\breve{L}$ obtained by replacing DG with $\breve{D}\breve{G}$ in L.

It should also be noted that $\breve{B} = B + O(h)$, and, in many cases, in the discrete version of the IBP, the approximate solution p (which is a result of using the simpler B instead of $\breve{B}$) in the discrete version of IBP, will not differ from those solutions $\breve{p}$ using $\breve{B}$ in an accurate, meaningful way. However, $\breve{B}$ should be kept for use in evaluating the boundary flux of $\breve{B}\breve{G}\breve{p}$.

The numerical solution $\breve{p}$ is obtained by solving the following single matrix equation, based upon $\breve{\mathbf{B}}$ instead of B:

$$(\alpha A + \beta \breve{\mathbf{B}}\breve{\mathbf{G}} - \breve{\mathbf{L}})\breve{p} = b. \tag{8.21}$$

When k is not a constant scalar, but rather a scalar-valued function of x, it can then be said that we are dealing with an isotropic but nonhomogeneous porous rock. In this case, the vector flux density is

$$-k(\mathbf{x})\nabla p(x) = \mathbf{v}, \tag{8.22}$$

so that the discrete version of IBP for any $\tilde{f}$, $\tilde{\mathbf{v}}$ (not related to p or $\breve{p}$) is

$$h < \hat{\breve{\mathbf{D}}}\tilde{\mathbf{v}}, \tilde{f} >_{\hat{Q}} + h < \breve{\mathbf{G}}\tilde{f}, \tilde{\mathbf{v}} >_P = < \breve{\mathbf{B}}\tilde{\mathbf{v}}, \tilde{f} >, \tag{8.23}$$

or

$$h < \hat{Q}\hat{\breve{\mathbf{D}}}\tilde{\mathbf{v}}, \tilde{f} > + h < P\breve{\mathbf{G}}\tilde{f}, \tilde{\mathbf{v}} > = < \breve{\mathbf{B}}\tilde{\mathbf{v}}, \tilde{f} >, \tag{8.24}$$

and, after renaming $\tilde{f} = \tilde{k}$ and $\tilde{\mathbf{v}} = -\breve{\mathbf{G}}\breve{p}$, our IBP becomes

$$h < \hat{Q}\hat{\breve{\mathbf{D}}}\breve{\mathbf{G}}\breve{p}, \tilde{k} > + h < (P\breve{\mathbf{G}})^T\breve{\mathbf{G}}\breve{p}, \tilde{k} > = < \breve{\mathbf{B}}\breve{\mathbf{G}}\breve{p}, \tilde{k} > . \tag{8.25}$$

It is important to notice that, in (8.22), $k(\mathbf{x})$ is a new name for $k(\mathbf{x})\lambda$, in order to have a physically meaningful Darcy velocity $\mathbf{v}$, and, of course, the relative permeabilities k_{rn} and k_{rw} appearing in the parameter λ are defined as quotients of k_n and k_w, with the absolute (scalar) permeability $k(\mathbf{x})$, depending upon position $\mathbf{x}$.

For the anisotropic case, important works have been conceived intending to explain the concept of relative permeability in anisotropic porous media and its implications in modeling fluid displacement. Specifically, [163] concludes that for each of the considered phases, the effective permeabilities $\mathscr{K}_w$ and $\mathscr{K}_n$ are also second-rank (symmetric) tensors, as can be rigorously shown by averaging the Navier–Stokes equations for a fluid that occupies only a part of the void space. Nonetheless, when defining the relative permeabilities in an anisotropic porous medium, as long as the definition is restricted to the principal x-, y-, and z-direction of fluid flow, the relative permeabilities (in each direction x, y, or z) can be defined for the i-th phase, as follows:

$$k_{ri,x}(S_e) \triangleq \frac{k_{i,x}(S_e)}{K_x},\tag{8.26}$$

$$k_{ri,y}(S_e) \triangleq \frac{k_{i,y}(S_e)}{K_y},\tag{8.27}$$

$$k_{ri,z}(S_e) \triangleq \frac{k_{i,z}(S_e)}{K_z},\tag{8.28}$$

where S_e denotes the **effective pressure**, defined by

$$S_e \triangleq \frac{(S_w - S_{wir})}{(1 - S_{wir})},\tag{8.29}$$

with S_{wir} denoting the **irreducible water saturation**; that is, the saturation at which the effective permeability of the wetting phase reduces to zero, due to the fact that it stops being a continuous phase within the void space.

In this case, the parameter λ is "absorbed" in the definition of "material tensor," which equals the product of λ and the original permeability tensor $\mathscr{K}$, a material tensor which is again renamed as $\mathscr{K}$.

Clearly, we can now compute the net boundary flux as $- < \breve{B}\breve{G}\breve{p}, \tilde{k} >$ or $- < B\breve{G}p, \breve{k} >$, when using the simpler boundary operator B.

Also, we must have $\tilde{\mathscr{K}} = DIAG(k(x_0), k(x_1), ..., k(x_n))$ and replace DG by $\breve{D}\tilde{\mathscr{K}}\breve{G}$ in L, to obtain $\breve{L}(\tilde{\mathscr{K}})$, thus considering nonhomogeneous porous rock.

Efficient numerical results using Castillo–Grone operators $\breve{D}$ and $\breve{G}$ have been obtained in [164], both for Poisson's equation (8.12) and the generalized equation $div(\mathscr{K} \cdot grad\ p) = -F$. The mimetic algorithms therein implemented are analyzed in one- and two-dimensional uniform staggered grids. In the case of one-dimension, algorithms for nonuniform grids have also been investigated.

For second-order mimetic discretization, Q is the identity matrix and the matrices P and $\breve{B}$ are given by

$$P = \begin{bmatrix} 3/8 & 0 & 0 & 0 & ... & ... & 0 \\ 0 & 9/8 & 0 & 0 & ... & ... & 0 \\ 0 & 0 & 1 & 0 & ... & ... & 0 \\ \vdots & ... & ... & \ddots & ... & ... & \vdots \\ 0 & ... & ... & 0 & 1 & 0 & 0 \\ 0 & ... & ... & 0 & 0 & 9/8 & 0 \\ 0 & ... & ... & 0 & 0 & 0 & 3/8 \end{bmatrix}$$

and

$$\check{\mathbf{B}} = \begin{bmatrix} -1 & 0 & 0 \dots & & & 0 \\ 1/8 & -1/8 & 0 & & & \vdots \\ -1/8 & 1/8 & 0 & & & \vdots \\ 0 & 0 & 0 & \ddots & 0 & 0 & 0 \\ \vdots & & & & 0 & -1/8 & 1/8 \\ \vdots & & & & 0 & 1/8 & -1/8 \\ 0 & & \dots & 0 & 0 & 1 \end{bmatrix}.$$

The corresponding mimetic operators $\check{\mathbf{D}}$ and $\check{\mathbf{G}}$ are given by

$$h\check{\mathbf{D}} = \begin{bmatrix} -1 & 1 & 0 & 0 & \dots & 0 \\ 0 & -1 & 1 & 0 & & \vdots \\ \vdots & & \ddots & \ddots & & \\ & & 0 & -1 & 1 & 0 \\ 0 & \dots & 0 & 0 & -1 & 1 \end{bmatrix}$$

and

$$h\check{\mathbf{G}} = \begin{bmatrix} -8/3 & 3 & -1/3 & 0 & \dots & & 0 \\ 0 & -1 & 1 & 0 & \dots & & 0 \\ \vdots & & & \ddots & & & \vdots \\ 0 & \dots & & 0 & -1 & 1 & 0 \\ 0 & \dots & & 0 & 1/3 & -3 & 8/3 \end{bmatrix}.$$

8.1.5 Arising Systems of Equations

In the 1-D uniform grid mimetic scheme for Poisson's equation and the given scalars α and β specifying the Robin's boundary condition, the matrix equation (8.21) appears in the following expanded forms, which are associated with using the simple B operator, and with $\check{\mathbf{B}}$, respectively. First, we present the system with simple B:

$$
\begin{bmatrix}
\left(\alpha + \frac{8\beta}{3h}\right) & -\frac{3\beta}{h} & \frac{\beta}{3h} & 0 & 0 & 0 & \cdots & & 0 \\
-8/3 & 4 & -4/3 & 0 & 0 & 0 & \cdots & & 0 \\
0 & -1 & 2 & -1 & 0 & 0 & \cdots & & 0 \\
0 & 0 & -1 & 2 & -1 & 0 & \cdots & & 0 \\
\vdots & & & \ddots & \ddots & \ddots & & & \vdots \\
0 & \cdots & & 0 & -1 & 2 & -1 & 0 & 0 \\
0 & \cdots & & 0 & 0 & -1 & 2 & -1 & 0 \\
0 & \cdots & & 0 & 0 & 0 & -4/3 & 4 & -8/3 \\
0 & \cdots & & 0 & 0 & 0 & \frac{\beta}{3h} & -\frac{3\beta}{h} & \left(\alpha + \frac{8\beta}{3h}\right)
\end{bmatrix}
p = \mathbf{F}. \quad (8.30)
$$

Secondly, consider the system with $\check{\mathbf{B}}$:

$$
\begin{bmatrix}
\left(\alpha + \frac{8\beta}{3h}\right) & -\frac{3\beta}{h} & \frac{\beta}{3h} & 0 & 0 & 0 & \cdots & & 0 \\
\left(\frac{-8-\beta h}{3}\right) & \left(\frac{\beta h+8}{2}\right) & \left(\frac{-8-\beta h}{6}\right) & 0 & 0 & 0 & \cdots & & 0 \\
\frac{\beta h}{3} & \left(\frac{-2-\beta h}{2}\right) & \left(\frac{\beta h+12}{6}\right) & -1 & 0 & 0 & \cdots & & 0 \\
0 & 0 & -1 & 2 & -1 & 0 & \cdots & & 0 \\
\vdots & & & \ddots & \ddots & \ddots & & & \vdots \\
0 & \cdots & & 0 & -1 & 2 & -1 & 0 & 0 \\
0 & \cdots & & 0 & 0 & -1 & \left(\frac{\beta h+12}{6}\right) & \left(\frac{-2-\beta h}{2}\right) & \frac{\beta h}{3} \\
0 & \cdots & & 0 & 0 & 0 & \left(\frac{-8-\beta h}{3}\right) & \left(\frac{\beta h+8}{2}\right) & \left(\frac{-8-\beta h}{6}\right) \\
0 & \cdots & & 0 & 0 & 0 & \frac{\beta}{3h} & -\frac{3\beta}{h} & \left(\alpha + \frac{8\beta}{3h}\right)
\end{bmatrix}
\check{p} = \mathbf{F}.
$$

$$(8.31)$$

For both systems, we denote

$$
\mathbf{F} = \begin{bmatrix}
l_b \\
h^2 F(x_{1/2}) \\
h^2 F(x_{1/2}) \\
\vdots \\
h^2 F(x_{n-3/2}) \\
h^2 F(x_{n-1/2}) \\
r_b
\end{bmatrix}
\quad (8.32)
$$

The main goal of Huy Khanh Vu's thesis [164] was to investigate the use of mimetic discretization for solving $(\alpha A + \beta B G - L(\mathcal{K}))p = -F$ on a 2-D uniform rectangular grid. More specifically, for use with the symmetric, positive-definite and second-order tensor matrix $\mathcal{K}$ (not simply diagonal).

Uniform grids are defined by the spatial steps Δx and Δy, which discretize $[x_L, x_U]$ and $[y_L, y_U]$, respectively. L and U refer to the lower and upper boundaries of $\Omega = (x_L, x_U) \times (y_L, y_U)$.

The continuum problem for the unknown pressure p is defined by the equation

$$div(\mathcal{K} \cdot grad\ p) = -f \text{ for all } (x, y) \in \Omega, \qquad (8.33)$$
$$\alpha p + \beta < grad\ p, \mathbf{n} > = \psi \text{ for all } (x, y) \in \partial\Omega,$$

where ψ is the given right-hand function at the boundary $\partial\Omega$ of the rectangle Ω. A typical example is worked out with a full tensor, defined as:

$$\mathcal{K} = \begin{bmatrix} 1 & 0.5 \\ 0.5 & 3 \end{bmatrix}. \qquad (8.34)$$

Usually, the material tensor coefficients are relatively small, ranging from 0.5 to 3 (numbers which are close to 1).

8.1.6   Example Problem and Results

Now, let us consider a test problem which was solved via the uniform cell centered method described in [165], for which the right-hand function f of the elliptic PDE equation (8.33) has been defined on $\Omega = (0, 1) \times (0, 1)$, as given by

$$
\begin{aligned}
f(x, y) = -\{ & [-\cos(x)(5 + 6.2\cos(y) - 7.1\cos(2y) + 8\cos(3y)) \\
& -4\cos(2x)(9 - 10\cos(y) + 1.1\cos(2y) + 12\cos(3y))] \\
& +3[2.1\cos(y) - 12.4\cos(2y) - 36\cos(3y) \\
& +\cos(x)(-6.2\cos(y) + 28.4\cos(2y) - 72\cos(3y)) \\
& +\cos(2x)(10\cos(y) - 4.4\cos(2y) - 108\cos(3y))] \\
& +[-\sin(x)(-6.2\sin(y) + 14.2\sin(2y) - 24\sin(3y)) \\
& -2\sin(2x)(10\sin(y) - 2.2\sin(2y) - 36\sin(3y))]\}.
\end{aligned}
$$

The following exact solution was used for f:

$$
\begin{aligned}
p(x, y) = 1 & - 2.1\cos(y) + 3.1\cos(2y) + 4\cos(3y) \qquad (8.35) \\
& + \cos(x)(5 + 6.2\cos(y) - 7.1\cos(2y) + 8\cos(3y)) \\
& + \cos(2x)(9 - 10\cos(y) + 1.1\cos(2y) + 12\cos(3y)).
\end{aligned}
$$

For testing purposes, the values of $p(x, y)$ and $< grad\ p, \mathbf{n} > (x, y)$ for all $(x, y) \in \partial\Omega$ can be obtained from equation (8.35) for both Dirichlet ($\beta = 0$) and Robin boundary conditions.

The results for the test problem using Dirichlet boundary conditions with $\alpha = 1$ are shown in Table 8.1.6; comparing computations from the mimetic method with the results of the cell-centered method given in [165]. Here, n is the number of cell center nodes in a 1-D uniform staggered grid, so that $x_i = i\Delta x$, $0 \le i \le n$, and $y_j = j\Delta y$, $0 \le j \le n$.

n	Mean Square			Max	
	Error (Mimetic)	Error (Cell-centered)	Order (Mimetic)	Error (Mimetic)	Order (Mimetic)
10	3.23E-02	9.65E-02		6.25E-02	
20	7.90E-03	2.44E-02	2.0316	1.63E-02	1.9390
40	2.00E-03	6.08E-03	1.9819	4.20E-03	1.9564
50	1.20E-03		2.2892	2.70E-03	1.9800
60	8.66E-04		1.7903	1.90E-03	1.9274

Table 8.1: Results for Dirichlet boundary conditions.

It should be noted that, overall, the mimetic method obtained the best result in terms of the mean square error (using both Dirichlet and Robin boundary conditions.)

8.1.7 Sample Problems

1. The following problem was taken from [55]. Solve the following:

$$-div(\mathcal{K}\,grad\,u) = F(x,y), \qquad (8.36)$$

on $(0,1)$, where

$$F(x,y) = -2(1 + x^2 + xy + y^2), \qquad (8.37)$$

considering Dirichlet boundary conditions. Let

$$\mathcal{K} = \begin{bmatrix} 2 & 1 \\ 1 & 2 \end{bmatrix}. \qquad (8.38)$$

Now, compare your solution with the following analytical solution:

$$u(x,y) = e^{xy}. \qquad (8.39)$$

2. The following problem was taken from [166]. Solve the following:

$$-div(\mathcal{K}\,grad\,u) = F(x,y), \qquad (8.40)$$

on $(0,1)$, where

$$F(x,y) = -[-22(y - y^2) - 26(x - x^2) + 18(1 - 2x)(1 - 2y)], \quad (8.41)$$

considering Dirichlet boundary conditions. Let

$$\mathcal{K} = \begin{bmatrix} 11 & 9 \\ 9 & 13 \end{bmatrix}. \tag{8.42}$$

Compare your solution against the following analytical solution:

$$u(x, y) = (x - x^2)(y - y^2). \tag{8.43}$$

3. Consider the continuity equation presented in Chapter 2. Similarly, consider **Darcy's law**, which relates the flux **u** of a homogeneous fluid with the effects of the pressure gradient ∇p in the reservoir, as follows:

$$\mathbf{u} = -\frac{1}{\mu}\mathcal{K}(\nabla p - \rho g \nabla z), \tag{8.44}$$

where μ is the fluid viscosity, ρ is the fluid's density, g is the magnitude of the gravitational acceleration, z is the vertical coordinate (positive upwards), and $\mathcal{K}$ is the absolute permeability tensor of the porous medium:

$$\mathcal{K} = \begin{bmatrix} K_x & 0 & 0 \\ 0 & K_y & 0 \\ 0 & 0 & K_z \end{bmatrix}, \tag{8.45}$$

where the components K_x, K_y, and K_z denote the permeability constants in the x-, y-, and z-directions, respectively. If $K_x = K_y = K_z$, the medium is said to be **isotropic**; otherwise, it is called **anisotropic**.

(a) Consider a generic conservation of mass equation for the density $\rho(\mathbf{x}, t)$ of the fluid, as follows:

$$\frac{d}{dt}\iiint_V \phi\rho(\mathbf{x}, t)dV = -\oiint_S \rho(\mathbf{x}, t)\mathbf{u}(\mathbf{x}, t) \cdot \mathbf{n}dS + \iiint_V q(\mathbf{x}, t)dV, \tag{8.46}$$

where **n** denotes the unit normal to the surface element dS, and $q(\mathbf{x}, t)$ denotes the mass flow rate per unit volume. What is the meaning of the previous statement? Can you describe the involved terms?

(b) Deduce that

$$\frac{\partial(\phi\rho)}{\partial t} = \nabla \cdot \frac{\rho}{\mu}\mathcal{K}(\nabla p - \rho g \nabla z) + q. \tag{8.47}$$

(c) Consider the definition of the compressibility of a fluid:

$$c \triangleq \frac{1}{\rho}\frac{\partial\rho}{\partial p}\bigg|_T, \tag{8.48}$$

Deduce the pressure equation

$$\nabla \cdot \frac{1}{\mu} \mathscr{K} (\nabla p - \rho_0 g \nabla z) + \frac{q}{\rho} = 0, \qquad (8.49)$$

which represents an elliptic partial differential equation for a single fluid phase in a porous medium.

8.2 Modeling Carbon Dioxide Geologic Sequestration

8.2.1 Introduction

The steady accumulation of greenhouse gases resulting from the combustion of fossil fuels has led to an increase in the amount of solar radiation trapped between the Earth and its atmosphere, leading to a rise in the temperature of both the Earth's atmosphere and its ocean systems. Many researchers believe that these continuing increases in temperature will lead to catastrophic changes in weather conditions around the globe. With carbon dioxide (CO_2) being the most abundant greenhouse gas, efforts are underway to reduce the amount of CO_2 entering the atmosphere.

Carbon dioxide capture, utilization, and sequestration (CCUS) is now considered a promising collection of technologies by which to assist in the mitigation of the environmental impact of greenhouse gases arising from the combustion of fossil fuels. More specifically, these technologies seek to separate the CO_2 from the flue gases being expelled by coal-fired power plants; compress the collected CO_2 to a supercritical phase ($ScCO_2$); and inject it into underground formations such as exhausted gas reservoirs or deep brine aquifers.

In this sense, in order to analyze the long-term behavior of the sequestered CO_2, water–rock interaction and reactive transport models are an invaluable resource [1]. Figure 8.1 shows this process schematically, including the need for the study of geochemical reactions following injection, which is depicted as an area of active research.

8.2.2 The Importance of Studying the Long-Term Behavior of Injected CO_2

As mentioned earlier, CCUS is a promising means of mitigating global warming. However, if a significant amount of CO_2 is to be sequestered in underground reservoirs, the geochemical implications must be analyzed first. This is particularly true in cases where there is large-scale pressure build-up in response to CO_2 injection, which may limit the dynamic storage capacity of suitable formations, due to the fact that overpressurization can fracture the caprock, causing CO_2 leakage and undesired induced seismicity [167]. The

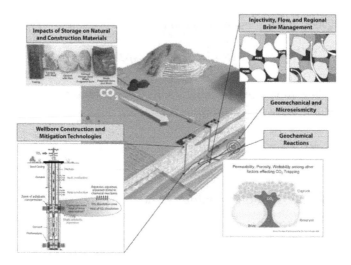

Figure 8.1: Conceptualization of the process of CO_2 capture and storage, highlighting areas of active research. Source: [1].

possibility of stored CO_2 leaking poses a significant risk, as is made clear by a frequently cited example which occurred on August 21, 1986.

The disaster began with the release of roughly one cubic kilometer of gaseous carbon dioxide into the atmosphere from the floor of Lake Nyos, in the hilly jungle terrain of western Africa. The plume looked like a white cloud as it spilled out of the lake and rolled into the surrounding valleys. By sunrise, more than 1,700 people and 3,200 animals had died of asphyxiation [168].

Still, the benefits of CCUS make it hard to ignore; particularly, as it has been proven that power plants equipped with CCUS technology produce approximately 80% to 90% less carbon than those without it, and CCUS also has the potential to reduce the cost of climate stabilization by 30% [168]. Further, it is believed that CO_2 can remain sequestered in such formations for up to 1,000 years (depending on the chemical and mechanical characteristics of the underground resident water and rock constituents).

8.2.3   Enhanced Oil Recovery and CCUS

One example of the potential economic benefits of using injected CO_2 has been established by an assortment of enhanced oil recovery (EOR) methods that were being utilized in hydrocarbon recovery [162].

Several methodologies for oil extraction have been broadly studied and implemented thus far. The simplest example is extraction by means of the extremely pronounced pressure gradient within the reservoir. This type of recovery is called **primary recovery**, and the pressures involved are high enough to force the resident fluids to flow through the porous medium and out of the wells with minimal pumping. Very often this method leaves 70% to 85% or more of the hydrocarbons inside the reservoir.

Secondary recovery techniques are based on the use of a fluid (such as water), that can be injected into some type of well. These techniques serve a dual purpose, as they are able to maintain high-reservoir pressures and flow rates while flooding the porous medium in order to physically displace some of the oil and push it toward the production well. Unfortunately, these techniques are not yet efficient enough, and significant amounts of hydrocarbons often remain in the reservoir (50% or more).

In order to recover more of the residual hydrocarbons, several enhanced recovery techniques involving complex chemical and thermal effects have been developed. These techniques form part of a variety of methods termed **tertiary recovery**. There are many different forms of tertiary EOR techniques. One method is based on the injection of gases like CO_2, which then mix with the resident hydrocarbon, resulting in a change to a single fluid phase. If complete mixing or miscibility is attained, the fluids will then flow together in one phase, thereby eliminating the distinction between possible phases, thus indicating that complete hydrocarbon recovery is (theoretically) possible. When a gas like CO_2 is used in the miscible displacement process, the phase change which allows it to achieve miscibility has the effect of producing a phase with a viscosity level somewhere between that of the gas and the oil. (The miscible phase flows more readily than oil.)

8.2.4 Water–Rock Interaction and Reactive Transport Modeling and Simulation

Researchers at the Computational Science Research Center at San Diego State University are developing a novel general water–rock interaction and reactive-transport simulator called *WebSymC*. This application allows a user-friendly specification of all required properties for scenarios in CO_2 injection and geologic sequestration; thereby allowing the simulation of the injection of CO_2-rich water into any configured geologic media (which also contains resident water), and analyzing the concentration of the involved species. A novel aspect of this software lies in its intuitive and educative web-based user interface, which was built using an Asynchronous JavaScript and XML (AJAX) methodology, and was designed to allow its users the ability to define any study case of interest.

Water-rock interaction and reactive transport modeling is an important tool for deciphering chemical and physical reactions which occur in sediments and rocks that are not accessible for monitoring purposes. The following form of the conservation of mass equation (which will be explained later in this section) is used for keeping track of the solutes' species mass in pore water (due to mass-transfer and reactions):

$$\frac{\partial e_\beta}{\partial t} = \phi \sum_{\alpha=1}^{Na} v_{\beta\alpha} \left(D_\alpha \nabla^2 c_\alpha - \nabla \cdot (c_\alpha \mathbf{u}) \right) - \mathcal{R}. \qquad (8.50)$$

An example study case of interest is studied by means of the simulation presented in [169], in which the formation of water was modeled as a solution comprised of 12 solutes (in concentrations similar to those found in the Frio Formation), with a geologic sandstone formation made up of an aggregation of six minerals, each with individual mineral volume fractions and grain radii similar to that of the sandstone found in the Frio Formation. The CO_2-rich injectant water was modeled as a mixture of the formation water with a 0.5 M solution of $CO_{2(aq)}$, with five times the molarity of iron and magnesium ions. The Fe^{++} ion acts as a tracer species in the injectant water and is used to designate the location of the advective sweep front, i.e., the location of the leading face of the injectant water, which encounters the reservoir's formation water. Investigations of the Frio Formation by Kharaka et al. (see [170]) found that, after 1,600 tons of CO_2 were injected, significant geochemical changes occurred when the $CO_{2(aq)}$ reached an observation well downstream of the injection well. Figures 8.2 and 8.3 depict the concentration of two of the species of interest for a period of five years.

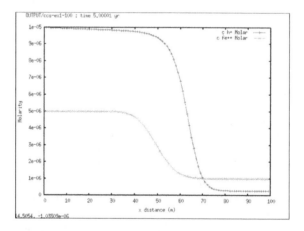

Figure 8.2: Results at 5 years of injection for the concentrations of Fe^{++} and H^+ with a grid resolution of 100 cells.

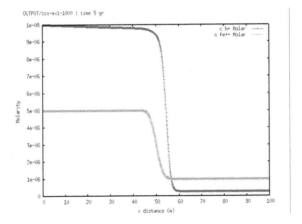

Figure 8.3: Results at 5 years of injection for the concentrations of Fe^{++} and H^+ with a grid resolution of 1,000 cells.

8.2.5 Potential Applications for Mimetic Discretization Methods

In order to illustrate the potential of mimetic discretization methods in this field, we must first consider a simplified initial-boundary value problem (which is used to analyze the long-term behavior of the concentration of aqueous CO_2 or $CO_{2(aq)}$).

Recall the extended form of Gauss' divergence theorem (see Appendix A):

$$\nabla \cdot (f\mathbf{v}) = \nabla f \cdot \mathbf{v} + f \nabla \cdot \mathbf{v}. \tag{8.51}$$

Now, let us consider the following form for the conservation of mass equation, which has been adapted for the study of flow through porous media:

$$\phi \frac{\partial c}{\partial t} = \phi D \nabla^2 c - \phi \nabla \cdot (c\mathbf{u}) - \mathscr{R} \Rightarrow$$

$$\frac{\partial c}{\partial t} = D \nabla^2 c - \nabla \cdot (c\mathbf{u}) - \frac{1}{\phi} \mathscr{R}.$$

From (8.51):

$$\frac{\partial c}{\partial t} = D \nabla^2 c - (\nabla c \cdot \mathbf{u} + c \nabla \cdot \mathbf{u}) - \frac{1}{\phi} \mathscr{R}. \tag{8.52}$$

Now, let us assume a near incompressible fluid: $\nabla \cdot \mathbf{u} = 0$.

$$\frac{\partial c}{\partial t} = D \nabla^2 c - \mathbf{u} \cdot \nabla c - \frac{1}{\phi} \mathscr{R}, \tag{8.53}$$

where:

1. c denotes the concentration of $CO_{2(aq)}$

2. D denotes the diffusivity coefficient for $CO_{2(aq)}$

3. $\mathbf{u}$ denotes the velocity field for the advected $CO_{2(aq)}$

4. ϕ denotes the porosity

5. $\mathscr{R}$ denotes the contribution due to chemical reactions

In cases of one spatial dimension, (8.52) becomes

$$\frac{\partial^2 c}{\partial x^2} - \frac{\mathbf{u}}{D}\frac{\partial c}{\partial x} - \frac{1}{D}\frac{\partial c}{\partial t} = \frac{\mathscr{R}}{\phi D}. \tag{8.54}$$

Equation (8.54) must be solved, subject to the following **initial condition**:

$$c(x,0) = 0.002 \text{ m.} \tag{8.55}$$

Notice that this initial value corresponds to some initial concentration of $CO_{2(aq)}$ in the resident water. Now, for the **boundary conditions**, consider the following:

$$\forall t \geq 0: \quad -\phi\frac{\partial c}{\partial x}(0,t) = 3 \text{ kg/s}, \tag{8.56}$$

$$\forall t \geq 0: \quad c(50 \text{ m}, t) = 0.002 \text{ molar.} \tag{8.57}$$

Typical parameters for our simplified model, considering $\mathscr{R} = 0$, are

1. $\phi = 0.32$

2. $D_{CO_2} = 7.8 \times 10^{-5} \text{ cm}^2/\text{s}$

3. $\mathbf{u} = 100 \text{ m/s}$

In one dimension, The Castillo–Grone mimetic operator $\breve{\mathbf{D}}$ allows us to accurately discretize the terms $\partial c/\partial x$ and $\partial^2 c/\partial x^2$, while the time-updating for $\tilde{c}(x_j, t_n)$ can be discretized by means of a standard Leapfrog scheme in time.

8.2.6   Initial Computational Implementation

In order to study the algorithmic approach, we first consider the following example of a one-dimensional diffusion-reaction equation in a purely mathematical scenario. This is a simplified version of the example introduced in Chapter 6. Recall our equation of interest:

$$-\nabla^2 f(x) = F(x), \tag{8.58}$$

where

$$F(x) = -\frac{\lambda \exp(\lambda x)}{\exp(x) - 1}, \tag{8.59}$$

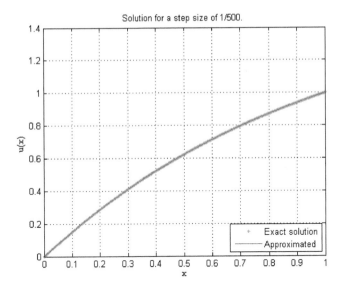

Figure 8.4: Attained solution of a diffusion-reaction problem by means of a second-order mimetic scheme.

will stand for our source term and accounts for any potential chemical reactions. Next, consider the following Robin's boundary conditions:

$$\alpha f(0) - \beta f'(0) = -1, \tag{8.60}$$
$$\alpha f(1) + \beta f'(1) = 0, \tag{8.61}$$

where

$$\lambda = -1 \tag{8.62}$$
$$\alpha = -\exp(\lambda) \tag{8.63}$$
$$\beta = -\frac{\exp(\lambda) - 1}{\lambda}. \tag{8.64}$$

A second-order mimetic discretization scheme yields the solution in Figure 8.4, with Figure 8.5 depicting the errors attained at the boundaries. Observe that the expected convergence at the boundaries is also second-order, as it was proposed for the interior nodes of the grid. Again, a more computationally intensive implementation of this example can be found in Chapter 6.

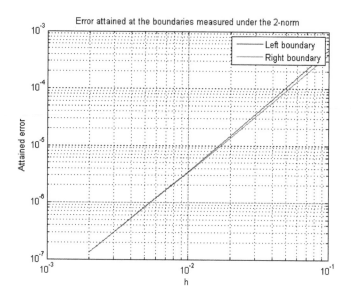

Figure 8.5: Attained error at the boundaries.

8.2.7   Sample Problems

1. Permeability measurements are very important in CCUS; more specifically, permeability studies of porous sandstone, a classic sedimentary rock composed mainly of sand-sized minerals or rock grains which is depicted in Figure 8.6, are of special importance, since this is the main compound of the lithologies in which supercritical CO_2 is injected.

Figure 8.6: A natural occurrence of sandstone.

Consider a 1-D sample of sandstone, such as in Figure 8.7, which depicts the sample, as well as the important properties to consider.

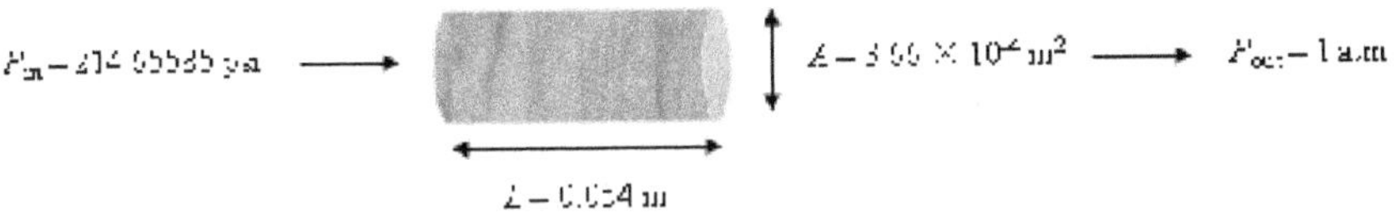

Figure 8.7: A 1-D sample of sandstone for permeability measurements in laboratory settings.

Considering Darcy's law, we are interested in computing the permeability of this sample of sandstone for flow of CO_2 at $T = 232$ °C, which has a related viscosity $\mu = 14.85 \times 10^{-6}$ Pa·s.

(a) Assuming that

$$P_{in}Q_{in} = P_{out}Q_{out} = \alpha \in \mathbb{R}, \qquad (8.65)$$

where Q stands for the fluid flow, and considering Darcy's law, deduce that the scalar permeability in the x-direction fulfills the following:

$$k = \frac{2P_{out}Q_{out}\mu L}{A(P_{in}^2 - P_{out}^2)}. \qquad (8.66)$$

(b) Considering that

$$Q_{out} = \frac{10 \times 10^{-6} \text{ m}^3}{16.97 \text{ s}} = 5.89 \times 10^{-7} \frac{\text{m}^3}{\text{s}}, \qquad (8.67)$$

Compute the permeability k in units of m^2.

(c) Compute the permeability in millidarcies (mD).

2. Can you duplicate the example problems in §8.2?

3. Can you extend Problem 2 to account for the advective flow? What about time evolution?

8.3 Maxwell's Equations

As seen in §2.3.2, time-dependent Maxwell's equations in free space ($\rho = k = 0$) can be written as

$$curl \ \mathbf{E} = -\frac{\partial \mathbf{B}}{\partial t}, \qquad (8.68)$$

$$curl \ \mathbf{H} = \frac{\partial \mathbf{D}}{\partial t}. \qquad (8.69)$$

The physical constants μ_0 and ρ_0 represent the permittivity and permeability of free space, respectively, with $\mathbf{B} = \mu_0\mathbf{H}$ and $\mathbf{D} = \epsilon_0\mathbf{E}$.

8.3.1 Background

In J.B. Runyan's M.Sc. Thesis (see [144]), the test problems used to compare mimetic methods based on the Castillo–Grone method with other finite differences methods are in 2-D, so that ($\partial/\partial z = 0$). In two dimensions, the six scalar equations corresponding to (8.68) and (8.69) are no longer interdependent and can be divided into two completely independent sets of interacting equations. These two sets are called the transverse electric (TE) mode for (E_x, E_y, H_z), and the transverse magnetic (TM) mode for (H_x, H_y, E_z). We have restricted our discussion to the TE mode in this text.

The application of computational methods to the solution of Maxwell's equations is called computational electrodynamics, or computational electromagnetics (CEM), and the first person to discretize Maxwell's equations was K.S. Yee (see [37]) in 1966. Yee's method was further developed by A. Taflove (see [171]) and others, and is called the finite difference time domain (FDTD) method.

Yee noticed [37] that, in cases where the plane is orthogonal to the original component, the time-derivative of each component is dependent upon the space-derivative of the components of the complementary field (**H** and **E** are regarded as complementary fields). Typically, we considering the TE mode

$$-\mu_0 \partial_t H_z = \partial_x E_y - \partial_y E_x, \tag{8.70}$$

$$\epsilon_0 \partial_t E_x = \partial_y H_z, \tag{8.71}$$

$$-\epsilon_0 \partial_t E_y = \partial_x H_z. \tag{8.72}$$

In TE mode, we have $H_x = H_y = 0$ and $E_z = 0$.

This means that it is not necessary to define each field component in a given discretized space, as it is sufficient to stagger the x, y, and z components of the two fields, in both space and time, as follows:

The **E** field components are defined for the integer time steps $n\Delta t$ with the **H** field components defined at half-integer time steps $(n + 1/2)\Delta t$. Moreover, the components affecting each other are now coplanar (i.e., a plane with fixed index k containing E_x, E_y, and H_z components, a plane with fixed index i contains E_y, E_z, and H_x components, and a plane with fixed index j contains E_z, E_x, and H_y components).

Yee's method [37] staggers the vector field components in both space and time, such that, if we consider the domain $\Omega = [0,1] \times [0,1] \times [0,1]$ to be divided into cubes with edge lengths of Δx, Δy, and Δz, then the electric field components are distributed tangentially along the edges of the cube. The magnetic field components, on the other hand, are distributed orthogonally among the faces of the cube. This arrangement is called the Yee cube and is illustrated in Figure 8.8.

Equation (8.70) tells us that the time evolution of H_z is directly proportional to the value of the z-component of *curl* **E**, etc.

Remembering that the face of the Yee cube is parallel to the $x - y$ plane (fixed index k) at time $(n - 1/2)\Delta t$, $n\Delta t$, and $(n + 1/2)\Delta t$, we can now

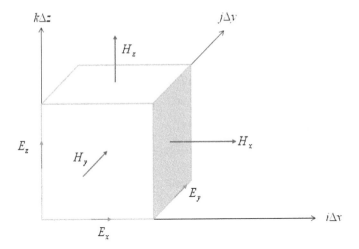

Figure 8.8: Yee's cube.

approximate the value of H_z at time $(n + 1/2)\Delta t$ in terms of the value of E_x and E_y at time $n\Delta t$ and H_z at $(n-1/2)\Delta t$, so that when we apply the second-order centered finite difference scheme to the time and space derivatives in equation (8.70), we get

$$-\mu_0 \frac{H_z^{n+1/2}(i + 1/2, j + 1/2, k) - H_z^{n-1/2}(i + 1/2, j + 1/2, k)}{\Delta t} =$$
$$\frac{E_y^n(i + 1, j + 1/2, k) - E_y^n(i, j + 1/2, k)}{\Delta x} -$$
$$\frac{E_x^n(i + 1/2, j + 1, k) - E_x^n(i + 1/2, j, k)}{\Delta y}. \qquad (8.73)$$

This equation can be solved for $H_z^{n+1/2}$ in terms of $H_z^{n-1/2}$ and values of E_x and E_y at time $n\Delta t$.

The staggered Yee scheme is a second-order accurate Leapfrog time integration, and is also second-order accurate in space.

8.3.2   Use of the Castillo–Grone Method

A discrete curl operator $\mathbf{C}_2$ is constructed from the second-order accurate Castillo–Grone (CG) 1-D divergence operator, so that the updated equation can be written as

$$H_z^{n+1/2} = H_z^{n-1/2} - \frac{\Delta t}{\mu_0} \mathbf{C}_2(E_x^n, E_y^n), \quad n = 1, 2, \dots \qquad (8.74)$$

After discretizing Equations (8.71) and (8.72), we obtain two similar formulas for updating E_x^{n+1} and E_y^{n+1}.

In TM mode, $E_x = E_y = 0$ and $H_z = 0$; and, as a result, (8.68) and (8.69) become

$$\epsilon_0 \partial_t E_z = \partial_x H_y - \partial_y H_x, \tag{8.75}$$
$$-\mu_0 \partial_t H_x = \partial_y E_z, \tag{8.76}$$
$$\mu_0 \partial_t H_y = \partial_x E_z. \tag{8.77}$$

The updated equation for E_z is now:

$$E_z^{n+1} = E_z^n + \frac{\Delta t}{\epsilon_0} \mathbf{C}_2(H_x^{n+1/2}, H_y^{n+1/2}), \; n = 0, 1, 2, ... \tag{8.78}$$

Now, recalling that the 2-D CG operator $\mathbf{C}$ is dependent upon the type of component matrix D we choose to construct it from, by choosing a k-th order accurate operator D, we expect the CG curl operator to also be k-th order accurate, as well, and we call this operator $(\mathbf{C}_k)$.

In his M.Sc. thesis, J.B. Runyan reviews sample problems in both one and two dimensions (for which the analytical solution is already known), and then compares two higher-order CG operators $\mathbf{C}_4$ and $\mathbf{C}_{4/343}$ (obtained from the 1-D operator, which is fourth order in the interior, but third order at the boundary). First, he compares it with the classic Yee scheme (using second-order accurate standard CFD first derivative operator), and second, he compares it with an ad hoc 4/343 scheme employed by Yefet and P.G. Petropoulos (see [172]), which they call the explicit (2, 4) scheme, already introduced by J. Fang (see [173]), and referring to the second-order accuracy in time and the fourth-order accuracy in space.

J.M. Hyman and M. Shashkov (see [52]) have provided support operator method (SOM) mimetic discretizations for the curl operator being applied to Maxwell's equations. On a uniform staggered grid, their method is equivalent to Yee's scheme (which is also equivalent to the second-order CG scheme). Hyman and Shashkov develop their method on nonuniform smooth and random grids. Note that the CG method developed for Maxwell's equations can also be applied to such grids, and can even be extended for use with higher-order operators; this is in contrast to the mimetic SOM operators provided by J.M. Hyman et al. (see [59]), which are inherently second-order.

8.3.3   A 1-D Example Problem

The 1-D sample problem under consideration in J.B. Runyan's thesis is a traveling wave in the unit interval along the x-axis, whose electric field component in the x direction is zero; as a result, the following initial boundary value problem (TE wave) is first defined in terms of the following **boundary conditions**:

$$E_y(0, t) = \sin(3\pi t), \; E_y(1, t) = -\sin(3\pi t). \tag{8.79}$$

Second, he considered the following **initial conditions**:

$$E_y(x, 0) = \sin(3\pi x), \; H_z(x, (1/2)\Delta t) = -\sin(3\pi(x + (1/2)\Delta t)). \tag{8.80}$$

The analytic solution is given by

$$E_y = \sin(3\pi(x+t)), \tag{8.81}$$
$$H_z = -\sin(3\pi(x+t)). \tag{8.82}$$

Empirically, it has been determined that, for $\Delta x = 1/40$, the CG 4/4 is stable for $\Delta t = 1/47$ and unstable for $\Delta t = 1/46$. These results are in agreement with the stability condition $\Delta t \le 6/7\Delta x$, as determined for the unbounded case provided by Yefet et al. (see [172]).

The results from the use of the operators C_4 and $C_{4/343}$ can now by compared with those of Yefet et al (see [172]) using first derivative matrices that are analogous to the matrices D_4 and D_{343}.

The L-2 norm of the error is defined as

$$\left(\sum_i \mathrm{err}_i^2 \Delta x \right)^{1/2}, \tag{8.83}$$

for each time step.

The maximum L-2 error in the time interval $[0, 4]$ is calculated for each of the three methods utilized for $\Delta x = 1/40$, as well as for the two temporal discretizations. Results are provided in Table 8.2.

h	Δt	CG:4/4	CG:4/343	Explicit (2,4)
$\frac{1}{40}$	$\frac{1}{720}$	2.703×10^{-5}	1.652×10^{-4}	1.652×10^{-4}
$\frac{1}{80}$	$\frac{1}{2330}$	8.368×10^{-7}	5.938×10^{-6}	6.361×10^{-6}
$\frac{1}{160}$	$\frac{1}{8640}$	2.645×10^{-8}	1.916×10^{-7}	2.090×10^{-7}

Table 8.2: For different temporal resolutions, the CG:4/343 outperforms the other two methods after calculating the maximum L-2 error in the time interval $[0, 4]$.

For the two given spatial discretizations of step size Δx, an optimal time step must be found for each of the three methods. Not only did the CG:4/4 scheme, with its optimal time step, provide the lowest overall error for each of the values of Δx considered $(1/40, 1/80, 1/160)$, but the optimal time step for the CG: 4/4 scheme was greater than that for each of the other two schemes.

8.3.4 A 2-D Example Problem

The 2-D sample problem in J.B. Runyan's thesis is also taken from Yefet et al. It models a transverse magnetic wave propagating along a rectangular waveguide with perfectly conducting walls (the tangential components of the electric field vanish, and the normal component of the magnetic field vanishes on the surface), with the analytic solution given by

$$E_z(x, y, t) = \sin(3\pi x - 5\pi t)\sin(4\pi y), \qquad (8.84)$$

$$H_y(x, y, t) = -\frac{3}{5}\sin(3\pi x - 5\pi t)\sin(4\pi y), \qquad (8.85)$$

$$H_x(x, y, t) = -\frac{4}{5}\cos(3\pi x - 5\pi t)\cos(4\pi y). \qquad (8.86)$$

Therefore, they consider the following initial-boundary value problem (TM Wave) with the following set of initial conditions:

$$E_z(x, y, 0) = \sin(3\pi x)\sin(4\pi y), \qquad (8.87)$$

$$H_y(x, y, (1/2)\Delta t) = -\frac{3}{5}\sin(3\pi x - (1/2)5\pi \Delta t)\sin(4\pi y) = H_y^{1/2}, \quad (8.88)$$

$$H_x(x, y, (1/2)\Delta t) = -\frac{4}{5}\cos(3\pi x - (1/2)5\pi \Delta t)\cos(4\pi y) = H_x^{1/2}. \quad (8.89)$$

A Dirichlet boundary condition on the electric field is prescribed for all $t \geq 0$, as follows:

$$\forall y \in [0, 1]: \ E_z(0, y, t) = -\sin(5\pi t)\sin(4\pi y), \qquad (8.90)$$

$$\forall y \in [0, 1]: \ E_z(1, y, t) = \sin(3\pi - 5\pi t)\sin(4\pi y), \qquad (8.91)$$

$$\forall x \in [0, 1]: \ E_z(x, 0, t) = 0, \qquad (8.92)$$

$$\forall x \in [0, 1]: \ E_z(x, 1, t) = 0. \qquad (8.93)$$

Since there is no z-dependence, field $\mathbf{E}$ is numerically divergence-free.

Also, it is easily verified that

$$\left(\frac{\partial H_x}{\partial x} + \frac{\partial H_y}{\partial y}\right)(x, y, (1/2)\Delta t) = 0, \qquad (8.94)$$

so that the magnetic field $\mathbf{H} = (H_x, H_y, H_z)$ is initially divergence-free, and we can expect it to remain numerically so. Numerical simulations confirm this expectation to a high order of accuracy (see [172]).

Considering the spatial discretizations $h = \Delta x = \Delta y = 1/20, 1/40$, and $1/80$ for each value of h and for each method, there is an optimal time step that yields minimal error, as in the 1-D case. Considering a time interval $[0, 10]$, the maximum of the total L-2 error is computed as in the 1-D case, and for each h, the best performance is achieved by the CG:4/4 scheme.

For $h = 1/40$, we present Table 8.3.4.

8.3.5   Sample Problems

1. Duplicate the results presented in §8.3.3.

h	Δt	CG:4/4	CG:4/343	Explicit (2,4)
$\frac{1}{20}$	$\frac{1}{400}$	1.1402×10^{-2}	11.4107×10^{-2}	1.4203×10^{-2}
$\frac{1}{40}$	$\frac{1}{800}$	8.9932×10^{-5}	1.0026×10^{-3}	1.0747×10^{-3}
$\frac{1}{80}$	$\frac{1}{2500}$	3.1440×10^{-6}	4.3693×10^{-5}	4.6662×10^{-5}

Table 8.3: The maximum error over the time interval $[0,10]$ is presented for each of the three methods for different temporal resolutions. We see that the smallest errors occur for the CG:4/4 method.

2. Duplicate the results presented in §8.3.4.

3. The following problem is extracted from [174]. Integrate Maxwell's equations. Consider $\epsilon_0 = 8.85 \times 10^{-12}$ and $\mu_0 = 1.2566 \times 10^{-6}$. Consider an infinite domain and a plane wave given by

$$\mathbf{E}(x,t) = \begin{bmatrix} 0 \\ \sqrt{\mu_0/\epsilon_0}g((t-(x+0.1)\sqrt{\epsilon_0\mu_0})10^9) \end{bmatrix}, \qquad (8.95)$$

$$H_z(x,t) = g((t-(x+0.1)\sqrt{\epsilon_0\mu_0})10^9), \qquad (8.96)$$

where the impulse g is defined as

$$g(s) = \begin{cases} [\exp(-10(s-1)^2) - \exp(-10)]/[1 - \exp(-10)], & 0 \le s \le 2 \\ 0 & \text{otherwise.} \end{cases}$$
$$(8.97)$$

Assume a free space media: $\mu = \mu_0$ and $\epsilon = \epsilon_0$. Let your numerical domain be an annulus with inner radius 0.1 m and outer radius 1.1 m, but take advantage of the symmetry of the problem by solving the problem in the half domain:

$$\Omega = \left\{ (x,y) \in (0.1 < \sqrt{x^2 + y^2} < 1.1) \times (y > 0) \right\}. \qquad (8.98)$$

Define the tangential component of $\mathbf{E}$ to be zero on all boundaries except the surface of the inner cylinder, where this tangential component should equal that of the incident wave. The initial conditions should correspond to the time when the incident wave (traveling from left to right) just arrives at the inner cylinder. Solve the problem in a uniform polar grid modeling Ω as defined.

8.4 Wave Propagation

In this section, we present some of the results obtained by O. Rojas (see [175]) and Rojas et al. (see [176] and [26]) concerning the dispersionless

propagation of elastic Rayleigh pulses in homogeneous media and rupture
propagation.

8.4.1   Problem Formulation

Consider an infinite elastic and isotropic half-plane with a horizontal x-axis
and a vertical positive downward z-axis (Figure 8.9).

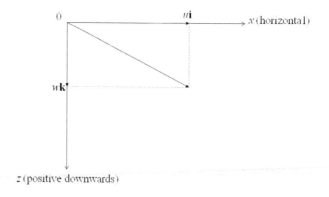

Figure 8.9: Half-space.

In this model, actual topography is idealized as planar, and a flat free
surface boundary condition is used to model the interaction of body waves
at this interface. The implementation of zero-traction conditions is tied to
the type of grid used in the discretization of the medium and wave-stress
fields, since the appropriate stress-tensor components must be zeroed at this
boundary (independent of their grid location).

The most popular grid for use in the finite-difference modeling of seismic
wave propagation is the staggered grid used in [177] to solve for the velocity-
stress formulations of the 2-D first-order elastodynamics equations of motion.
This grid has also been utilized in 3-D applications by various authors, and will
be hereafter referred to as the "standard staggered grid" (SSG). Alternately,
[178] employed a "rotated standard grid" (RSG) to simulate wave propagation
in 2-D elastic media containing both crack and free surfaces.

Due to the fact that higher-order approximations either become unstable
(SSG implementations) or fluctuate strongly (RSG implementations), only
second-order discretization of the equations of motion are used to locally compute the wave field.

To attain free surface implementations of higher accuracy on a SSG, consider utilizing the 2-D stress imaging formulations introduced in [179] (their
performance has also been studied in 3-D displacement stress formulations

(see [180])). Due to the staggered distribution of wave and stress fields on the grid, two different implementations of the zero-traction conditions are possible. In the first, the free surface is placed along the grid plane, where it undergoes normal stresses and horizontal velocity-displacement. In the second scenario, it shares the same grid plane as both the (xz, yz) stress tensor components and the (vertical, transversal) velocities-displacements. In [180], the first alternative is called H-formulation, and the second W-formulation (we hereafter adopt both of these nominations).

In [181], a Gaussian-shaped source buried two kilometers deep in a uniform half-space was used, with approximately six points per shear wavelength being allowed by the grid spacing. The researchers concluded that, once the horizontal velocity had been interpolated by simple averaging, the W-formulation produced more precise wave fields along the free surface, and also modeled the Rayleigh wave propagation more accurately.

Through use of the adjusted finite difference formulation provided by [180], it was discovered that, after testing both $H-$ and W-formulations in the experiments with a Gabor wavelet, the phases of the numerical waves were better reproduced by the H-formulation, while the W-formulation yielded better amplitudes.

8.4.2 Methodology

O. Rojas (see [176]) has proposed four staggered-grid solvers of the elastodynamic wave equations in 2-D media, including a flat free surface boundary. The full displacement vector $(u, 0, w)$ is obtained at this boundary, and fictitious gridlines are avoided. First of all, both the RSG and SSG grids have been enhanced by the inclusion of a new set of **compound nodes** (also called **displacement-stress nodes**) along the free surface, allowing the one-sided discretization of the zero-stress conditions and the computation of the displacement vector. He found that the set of mimetic finite-difference operators first proposed by [142] became convenient for this purpose because it comprises one-sided and centered staggered differentiators with multiple orders of accuracy.

O. Rojas found that the set of mimetic finite-difference operators first proposed by [142] was well suited for this purpose, as it was comprised of both one-sided and centered staggered differentiators with multiple orders of accuracy.

We shall briefly present the fourth-order algorithm (MCSG), designed on a staggered grid previously used in solving diffusion phenomena [51], [182] and Maxwell's equations [52].

The propagation of elastic waves in 2-D media (half-plane) is described by the following elastodynamic wave equations:

$$\rho \frac{\partial^2 u}{\partial t^2} = \frac{\partial}{\partial x}[\tau_{xx}] + \frac{\partial}{\partial z}[\tau_{xz}], \tag{8.99}$$

$$\rho \frac{\partial^2 w}{\partial t^2} = \frac{\partial}{\partial x}[\tau_{xz}] + \frac{\partial}{\partial z}[\tau_{zz}], \tag{8.100}$$

$$\tau_{xx} = (\lambda + 2\mu)\frac{\partial u}{\partial x} + \lambda \frac{\partial w}{\partial z}, \tag{8.101}$$

$$\tau_{zz} = (\lambda + 2\mu)\frac{\partial w}{\partial z} + \lambda \frac{\partial u}{\partial x}, \tag{8.102}$$

$$\tau_{xz} = \mu(\frac{\partial u}{\partial z} + \frac{\partial w}{\partial x}). \tag{8.103}$$

The last three equations represent the linear stress–strain relationship for elastic media, also known as **Hooke's law**. The mass density $\rho(x, z)$ and the Lamé parameters $\lambda(x, z)$ and $\mu(x, z)$ constitute the parameters of this model and determine the velocity of the compressional (P) wave velocity α, and the shear (S) wave velocity β through the relations:

$$\alpha = \sqrt{\frac{\lambda + 2\mu}{\rho}}, \tag{8.104}$$

$$\beta = \sqrt{\frac{\mu}{\rho}}. \tag{8.105}$$

In geophysics, the abovementioned system of equations is referred to as the **P-SV wave propagation problem.**

A Neumann-type boundary condition is imposed at $z = 0$ to model a free surface by vanishing the normal and tangential stresses, i.e., by setting $\tau_{xz} = \tau_{zz} = 0$.

An explosive point given by the Gaussian pulse $f(t) = \exp(-\delta(t - t_0)^2)$ is used for numerical experiments because of its simple implementation.

In cases where a source is located at an interior point, $f(t)$ is added to both normal stresses τ_{xx} and τ_{zz}, while a vertical point force applied at the free surface (Lamé's problem) is modeled by incrementing only τ_{zz} by $f(t)$ (see [183], citearticle-shuo).

Nonradiation boundary conditions along the other three edges of the computational domain (which are necessarily finite) have also been considered, with the simulation time being adjusted to avoid boundary reflections.

Now, consider the classical PS-V problem, chosen for its variety of interesting features, including the dispersionless propagation of Rayleigh pulses in homogeneous media with a known analytical solution.

The normal stresses τ_{xx}, τ_{zz} are defined at the same grid point, implying a unique way of updating the stress field as a result of source contribution.

Consider a classical rectangular grid, where both components of the displacement vector (u, w) are defined at every cell center, as well as along the free surface $z = 0$. The treatment of the traction-free boundary conditions yields the value (u, w) on the top face of boundary cells at $z = 0$. Using the spatial steps $(\Delta x, \Delta z)$ and the pair of indices (i, j), $1 \leq i \leq N_x$, $1 \leq j \leq N_z$, every cell is denoted by a tuple (i, j), with $(1, 1)$ located at the upper left corner.

For simplicity's sake, the discrete analog (u_{ij}, w_{ij}) corresponds to (u, w) at $((i - \frac{1}{2})\Delta x, (j - \frac{1}{2})\Delta z)$, which is the position of the center point of the cell (i, j). Finally, as will be explained later, the application of the free surface boundary condition leads to the boundary values (u_{i0}, w_{i0}), $1 \leq i \leq N_x$, with a physical location of $((i - \frac{1}{2})\Delta x, 0)$.

This simple, classical grid does not allow full computing of any of the components of the strain tensor (u_x, w_x, u_z, w_z) at any grid position. In fact, u_x and w_x could be computed at the center of both left and right faces of any cell, but u_z and w_z cannot be calculated at these locations. On the other hand, u_z and w_z are available at the center of the top and bottom faces, but u_x and w_x are not. To overcome this liability, fourth-order Lagrange interpolation of the closest values is used to approximate the unavailable quantity of each face, and then strain and stress tensors are conveniently defined. Approximations to x-derivatives and z-derivatives computed by finite difference formulas have no-hat symbols u_x, w_x, u_z, w_z, while hat symbols are used to denote approximation by interpolating, $\hat{u}_x, \hat{w}_x, \hat{u}_z, \hat{w}_z$.

The right chart in Figure 8.10 depicts the computation cell with displacement, stress, and material properties location.

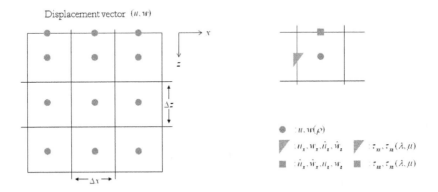

Figure 8.10: Distribution of displacement vectors along the classical staggered grid used by MCSG (left chart). Both components u and w are defined at every cell center, and also along the free surface. Locations of partial derivatives u_x, w_x, u_z, w_z calculated by staggered differentiation, as well as approximations $\hat{u}_x, \hat{w}_x, \hat{u}_z, \hat{w}_z$ given by interpolation, are illustrated in the right chart of this figure.

Boundary values (u_{i0}, w_{i0}), $1 \leq i \leq N_x$ at $z = 0$ are calculated by solving the zero-traction conditions and using staggered differentiators along every $x = i\Delta_x$ gridline and nodal differentiators along $z = 0$. In particular, MCSG was implemented using fourth-order operators G_{4-4-4} and N_{4-4-4}. The subindex 4-4-4 in operators is used to denote the fourth-order of accuracy at the interior and boundary points. N_{4-4-4} is the uniformly fourth-order approximation to the first derivative on the equally-spaced nodal grid proposed by [142]. Another set of mimetic operators, D_{2-4-2} and G_{2-4-2}, where the fourth-order of accuracy has been kept at the interior, but second-order formulas are given at the boundary points, are useful in coping with instabilities.

8.4.3 Application of the Castillo–Grone Method

Let us define $U_j = (u_{1j}, ..., u_{Nxj})^T$ and $W_j = (W_{1j}, ..., W_{Nxj})^T$ for $j = 1, ..., 6$. In addition, consider the matrix $I_\gamma = \text{DIAG}\,(\gamma_{10}, ..., \gamma_{Nx0})$ given by values of γ at every boundary gridpoint. Thus, $\tau_{xz} = \tau_{zz} = 0$ could be discretized as

$$(u_z + w_x)\,|_{0,j} = 0 \Rightarrow \frac{1}{\Delta_z}\sum_{j=1}^{6} g_{1j} U_{j-1} + \frac{1}{\Delta x} N_{4-4-4} W_0 = 0, \quad (8.106)$$

$$(\gamma u_x + w_z)\,|_{0,j} = 0 \Rightarrow \frac{1}{\Delta x} I_\gamma N_{4-4-4} U_0 + \frac{1}{\Delta_z}\sum_{j=1}^{6} g_{1j} w_{j-1} = 0. \quad (8.107)$$

Here, g_{1j} denote elements of the Castillo–Grone gradient operator $\check{\mathbf{G}}$. By rewriting the equations above, we get

$$U_0 + k N_{4-4-4} W_0 = -\sum_{j=2}^{6}(g_{1j}/g_{11}) U_{j-1} = b, \quad (8.108)$$

$$W_0 + k I_\gamma N_{4-4-4} U_0 = -\sum_{j=2}^{6}(g_{1j}/g_{11}) W_{j-1} = c, \quad (8.109)$$

where $k = \Delta z/(\Delta x g_{11})$. Systems (8.108) and (8.109) are easily decoupled, leading to:

$$(k I_\gamma N_{4-4-4}^2 - I) W_0 = (k I_\gamma N_{4-4-4}) b - c. \quad (8.110)$$

In cases where U_j and W_j for $j = 2$ are known, we can solve for W_0 first, since we have a system of N_x linear equations to determine it according to boundary conditions and discrete values of the wave field at the interior.

Thefore, we basically now have a three-step MCSG algorithm (see Algorithm 5).

The efficiency of this algorithm has been tested against the exact solution of the classical Lamb's problem; that is, the response of an elastic half-plane due to the application of a vertical point force $f(t)$ at $(x_0, z = 0)$.

Algorithm 5 A three-step MCSG algorithm.

STEP 1 (Interior): Solve the elastodynamic wave equation to get (u, w) for $t = (n+1)\Delta t$ at every interior node. Accuracy of fourth-order in space and second-order in time. Then compute vectors b and c.

STEP 2 (Free Surface): Solve (8.110) for W_0 at $t = (n+1)\Delta t$ using backward/forward substitution.

STEP 3 (Free Surface): Calculate U_0 for $t = (n+1)\Delta t$ from Equation (8.108), using W_0.

Algorithm 5 uses 6 points ($h = 12$ m) per minimum S-wavelength (λ_s^{min}), as opposed to the 10 points ($h = 6$ m) per (λ_s^{min}) employed by second-order algorithms. The time step chosen was $\Delta t = 0.5(\Delta x/\alpha)$.

The homogeneous half-space was such that $\rho = 2500\ kg/m^3$ and $\beta = 1500\ m/s$, filled with a Poisson solid ($\alpha = \sqrt{3}\beta$).

Low dispersive modeling of surface waves propagating along a flat free surface is achieved thanks to the incorporation of compound nodes at this boundary, and the high-order, one-sided differentiation of wave fields.

8.4.4 Attained Results

From the three-parametric families of fourth-order accurate $\check{G}$ and $\check{D}$ operators exhibited by the Castillo–Grone Method (see [142]), numerical tests were performed with $(\alpha, \beta, \gamma) = (\alpha^1, \beta^1, \gamma^1) = (0, 0, -1/24)$, so that the actual matrices used to run the tests were

$$G_{4-4-4} = \begin{bmatrix} -\frac{1775}{528} & \frac{1790}{407} & -\frac{2107}{1415} & \frac{1496}{2707} & -\frac{272}{2655} & \frac{25}{9768} & 0 \cdots \\ \frac{16}{105} & -\frac{31}{24} & \frac{29}{24} & -\frac{3}{40} & \frac{1}{168} & 0 & 0 \cdots \\ 0 & \frac{1}{24} & -\frac{27}{24} & \frac{27}{24} & -\frac{1}{24} & 0 & 0 \cdots \\ 0 & 0 & \frac{1}{24} & -\frac{27}{24} & \frac{27}{24} & -\frac{1}{24} & 0 \cdots \end{bmatrix}, \tag{8.111}$$

and

$$D_{4-4-4} = \begin{bmatrix} -\frac{1045}{1142} & \frac{909}{1298} & \frac{201}{514} & -\frac{1165}{5192} & \frac{129}{2596} & -\frac{25}{15576} & 0 \cdots \\ \frac{1}{24} & -\frac{27}{24} & \frac{27}{24} & -\frac{1}{24} & 0 & 0 & 0 \cdots \\ 0 & \frac{1}{24} & -\frac{27}{24} & \frac{27}{24} & -\frac{1}{24} & 0 & 0 \cdots \end{bmatrix}. \tag{8.112}$$

The next set of mimetic operators improve the stability:

$$D_{2-4-2} = \begin{bmatrix} -1 & 1 & 0 & 0 & 0 & 0 & 0 \cdots \\ \frac{1}{23} & -\frac{26}{23} & \frac{26}{23} & -\frac{1}{23} & 0 & 0 & 0 \cdots \\ 0 & \frac{1}{24} & -\frac{27}{24} & \frac{27}{24} & -\frac{1}{24} & 0 & 0 \cdots \\ 0 & 0 & \frac{1}{24} & -\frac{27}{24} & \frac{27}{24} & -\frac{1}{24} & 0 \cdots \end{bmatrix}, \tag{8.113}$$

$$G_{2-4-2} = \begin{bmatrix} -\frac{8}{3} & 3 & -\frac{1}{3} & 0 & 0 & 0 & 0 \cdots \\ \frac{4}{39} & -\frac{31}{26} & \frac{44}{39} & -\frac{1}{26} & 0 & 0 & 0 \cdots \\ 0 & \frac{1}{24} & -\frac{27}{24} & \frac{27}{24} & -\frac{1}{24} & 0 & 0 \cdots \\ 0 & 0 & \frac{1}{24} & -\frac{27}{24} & \frac{27}{24} & -\frac{1}{24} & 0 \cdots \end{bmatrix}, \tag{8.114}$$

and

$$N_{4-4-4} = \begin{bmatrix} -\frac{33989}{13640} & \frac{49453}{8184} & -\frac{28993}{4092} & \frac{7391}{1364} & -\frac{18763}{8184} & \frac{16717}{40920} & 0 \cdots \\ -\frac{1}{4} & -\frac{5}{6} & \frac{3}{2} & -\frac{1}{2} & \frac{1}{12} & 0 & 0 \cdots \\ \frac{1}{12} & -\frac{287}{348} & \frac{55}{87} & -\frac{49}{174} & \frac{191}{348} & -\frac{55}{348} & 0 \cdots \\ 0 & \frac{1}{12} & -\frac{2}{3} & 0 & \frac{2}{3} & -\frac{1}{12} & 0 \cdots \end{bmatrix}. \tag{8.115}$$

The numerical stability and grid dispersion of RSG finite-difference schemes were studied by [178] for the case of an infinite, elastic, and homogeneous medium. Different orders of spatial discretization and second-order in time were considered, and they proposed the following relation as the Von Neumann stability condition for an even order-spatial accurate scheme:

$$\frac{\alpha \Delta t}{h} \leq \frac{1}{\sqrt{2} \sum |C_A|} \tag{8.116}$$

In the above inequality, the scalars C_k are the coefficients of the stencil for central differentiation, and can be used to constrain the time step Δt.

An experiment similar to the one proposed by [184] was performed, with a vertical point force given by a narrow-banded Gaussian-type pulse, given by $f(t) = \exp[-500(t - 0.25)^2]$, with an approximate frequency spectrum of $[0, 20]$ Hz.

It was observed that low dispersive modeling of surface waves propagating along a flat free surface had been achieved, thanks to the incorporation of compound nodes at this boundary, and the high-order discretization of the wave fields.

The one-sided, high-order mimetic differentiation can be exploited to model alternative and challenging boundary conditions of the Neumann type, such as fault jump conditions for frictional sliding of contiguous elastic plates. Also, the effect of real topography on surface and body waves could be studied by adapting the mimetic algorithms to non-Cartesian grids.

Although, in Lamb's experiment the signal recorded by a superficial station is mainly the Rayleigh pulse, a basic approach has been introduced to isolate this pulse from body-wave arrivals. Letting x_3 denote the surface position of receiver R_3, then Figure 8.11 depicts both the analytical time series $u(x_3, 0, t)$ and $w(x_3, 0, t)$ at R_3. The Rayleigh wave arrival time TR;$R_3 = 11.31$ s, and the S-wave arrival time TS;$R_3 = 10.40$ s, exhibit a natural separation of pulses with a time gap of approximately 0.7 seconds. Comparing these two signals, $u(x_3, 0, t)$ reaches very low values during this time gap, but $w(x_3, 0, t)$ does not present this behavior. Thus, the first basic idea in the dispersion analysis of the numerical Rayleigh wave is to define a cut-off time TC;R_3 to truncate the time series of $u(x_3, 0, t)$, and keep records that only describe the Rayleigh pulse, TS;$R_3 < TC;R_3 < TR;R_3$. A simple definition is

$$TC; R_3 \triangleq \min_{TS;R_3 < t < TR;R_3} |u(x_3, 0, t)|. \tag{8.117}$$

The reason for considering the magnitude of $u(x_3, 0, t)$ is clearly exposed by the seismogram produced by MCSG. It shows how low-magnitude oscillations reach R_3 earlier than the important part of the Rayleigh pulse, illustrating a typical effect of numerical dispersion in fourth-order finite difference solutions, in which high-frequency components travel faster than low-frequency ones. The simulation time should be long enough to allow most of these low-amplitude oscillations to reach and be recorded at R_3.

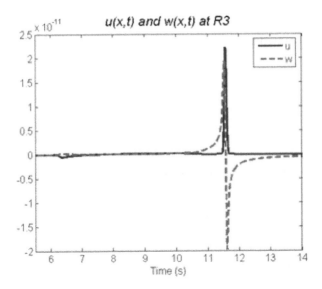

Figure 8.11: Horizontal (u) and vertical (w) analytical displacements at R_3. P-wave arrival time is 6.0 s, S-wave arrival time ($T_{S;R_3}$) $=10.40$ s, and Rayleigh wave arrival time ($T_{R;R_3}$) $=11.31$ s.

8.4.5   Sample Problems

1. Solve the following hyperbolic equations in (-1,1):

$$p_t + Ku_x = 0, \tag{8.118}$$

$$u_t + \frac{1}{\rho}p_x = 0. \tag{8.119}$$

Solve for a given bulk modulus of compressibility $K = 0.25$. In the latter, u stands for the velocity, for which you have to specify a reasonable numerical value, as well as for p, which stands for pressure. Assume the following initial conditions:

$$p(x,0) = \frac{1}{2}\exp(-80x^2) + S(x) \tag{8.120}$$

$$u(x,0) = 0 \tag{8.121}$$

with $S(x) = 0.5$ for $-0.3 < x < -0.1$ and $S(x) = 0$. Propose boundary conditions that model a solid wall.

2. Solve the following hyperbolic equations in (0,1):

$$\eta_t + Hu_x = 0 \tag{8.122}$$

$$u_t + g\eta_x = 0, \tag{8.123}$$

where g denotes the gravitational acceleration. Solve for the given depth $D = 10$. In the latter, u stands for the velocity, for which you have to specify a reasonable numerical value, as well as for η, which stands for a specific level perturbation. Let $\eta(x,0) = u(x,0)$ be the initial conditions, but propose boundary conditions that model both an oscillating and a solid wall.

3. Consider the following equation for the displacement vector $\mathbf{u}(x,t)$:

$$\rho\frac{\partial^2}{\partial t^2} = \nabla \cdot (\tau + \mathbf{m}), \tag{8.124}$$

where the stress–strain relations define the stress tensor, as follows:

$$\tau = \lambda(div\ \mathbf{u})\mathbf{I} + \mu(\nabla\mathbf{u} + \nabla\mathbf{u}^T). \tag{8.125}$$

Specify reasonable numerical values for the parameters, but assume a free-surface boundary condition of the form:

$$\tau = \mathbf{n} = \mathbf{0}. \tag{8.126}$$

8.5 Geophysical Flow

8.5.1 Equation of Motion in the Atmosphere and Oceans

The widely used mathematical equation that describes the motion of fluids is called the Navier–Stokes (NS) equation. This equation is named after Claude-Louis Navier (1785–1836, French engineer and physicist) and George Gabriel Stokes (1819–1903, British mathematician and physicist), who derived this equation by applying Newton's second law of motion to fluid flow. The NS equation can be used to simulate flow around a car, in a pipe, atmospheric flows, ocean currents, and even the flow of blood within the human body. However, in this section we focus solely on geophysical flows, including atmospheric and oceanic flows. In this particular case, the NS equation is written as

$$\underbrace{\frac{\partial u_i}{\partial t}}_{I} + \underbrace{u_j \frac{\partial u_i}{\partial x_j}}_{II} + \underbrace{\frac{\partial p}{\partial x_i}}_{III} - \underbrace{\frac{1}{Re}\frac{\partial^2 u_i}{\partial x_j^2}}_{IV} - \underbrace{\frac{1}{Ro}\varepsilon_{ij3}u_j}_{V} + \underbrace{\frac{1}{Fr^2}\beta\delta_{i3}}_{VI} = 0, \qquad (8.127)$$

where:

- Term I is the storage of the momentum

- Term II is the advection term

- Term III is the pressure effect

- Term IV is the diffusion

- Term V is the Coriolis force, due to the Earth's rotation

- Term VI is the buoyancy

Equation (8.127) uses Einstein's indexing rule, in which one has to sum over the repeated index, and is written in nondimensional form. (This means all the variables are nondimensional.) This is done by choosing an appropriate length scale (L^*), a velocity scale (U^*), and a reference density (ρ^*). Then, the variables are nondimensionalized, as follows:

$$(x_i, u_i, t, p) = \left(\frac{x_i^*}{L^*}, \frac{u_i^*}{U^*}, \frac{t^* U^*}{L^*}, \frac{p*}{\rho^*(U^*)^2} \right). \qquad (8.128)$$

There are a number of nondimensional variables used in Equation (8.127). The first is called the **Reynolds number**, which is the ratio of the inertial forces to the viscous forces

$$Re = \frac{U^* L^*}{\nu}, \qquad (8.129)$$

where ν is the kinematic viscosity, defined as

$$\nu = \frac{\mu^*}{\rho^*}. \tag{8.130}$$

In the above case, μ^* is the **viscosity** (a property of the fluid that measures its resistance to the deformation due to the tensile or shear stresses). Seawater at 20 °C with salinity of 35 g·kg^{-1}, has viscosity of 1.08×10^{-3} kg·m^{-1}·s^{-1}. Considering the large length scales of the atmospheric and oceanic flows, and taking into account the properties of seawater and the atmosphere, one can conclude that the Reynolds number in geophysical flows is much larger than 10^6. Hence, it is very common to ignore the diffusion term in the NS equation, while simulating atmospheric and oceanic flows.

The next nondimensional number in Equation (8.127) is called the **Rossby number**, which takes into account the Earth's rotation, and is defined as

$$Ro = \frac{U^*}{fL^*}, \tag{8.131}$$

where

$$f = 2\omega \sin \phi. \tag{8.132}$$

Note that, in the abovementioned equation, ϕ is the latitude and $\omega = 0.725 \times 10^{-4}$ s^{-1}. The last nondimensional parameter is called the **Froude number**, which is defined as

$$Fr = \frac{U*}{\sqrt{gL^*}}, \tag{8.133}$$

where g is the Earth's acceleration of gravity. And finally, the **buoyancy ratio**, β, is defined as

$$\beta = \frac{\rho'}{\rho^*}, \tag{8.134}$$

where $\rho' = \rho - \rho^*$ is the deviation from the reference density. Density is defined by the equation of the state. In oceans, the density is a function of the pressure, salinity, and temperature. The UNESCO equation of state (EOS) is widely used. In atmospheric flows, the density is a function of the pressure, temperature, and water vapor content. Using the perfect gas law, the buoyancy ratio can be written as

$$\beta = \frac{\theta'}{\theta} \tag{8.135}$$

The remaining two symbols in Equation (8.127) are defined as

$$\varepsilon_{ijk} = \begin{cases} +1 & ijk = 123, 231, 312 \\ -1 & ijk = 321, 213, 132 \\ 0 & \text{otherwise}, \end{cases} \tag{8.136}$$

and

$$\delta_{ij} = \begin{cases} 1 & i = j \\ 0 & \text{otherwise}. \end{cases} \tag{8.137}$$

8.5.2 Boussinesq Approximation and Incompressible Equations

Under the assumption of incompressibility and constant viscosity, μ^*, the shear stress tensor term becomes $\mu \nabla^2 \vec{v}$, where ∇^2 is the vector Laplacian.

Equation (8.127) is widely known as the incompressible Navier-Stokes equation, and can be used for both atmospheric and oceanic flows. While the water is considered to be incompressible, one might question the validity of the abovementioned equation when simulating atmospheric flows. A more accurate look at Equation (8.127) reveals that any change in density is ignored, except when it is multiplied by the Earth's acceleration g. Therefore, one might question the validity of the abovementioned equation even in oceanic simulations. However, even when dealing with a compressible fluid, such as the air at room temperature, at low Mach numbers (even up to Mach 0.3), the incompressibility flow assumption typically holds well.

As a matter of fact, Equation (8.127) is called the incompressible Navier-Stokes equation with Boussinesq approximation. Joseph Valentin Boussinesq (1842–1929, French mathematician and physicist) showed that, under certain conditions, which are widely satisfied in geophysical flows, it is possible to ignore the changes in the density, except when it is multiplied by the Earth's acceleration, and the constant density is equivalent to a null divergence of velocity.

His assumptions considerably simplify the equation of motion. However, in order to be able to use Boussinesq's approximation, the following conditions must first be satisfied:

- The velocity must be much smaller than the speed of sound (c)

- The phase speed must also be much smaller than c

- The scale of motion must be smaller than c^2/g

These conditions are generally satisfied for atmospheric and oceanic flows.

8.5.3 Shallow Water Equations

The horizontal length scale is much larger than the vertical scale, in both the atmosphere and the oceans. Therefore, in large-scale models, the vertical velocity is much smaller than the velocity's two horizontal components. As a result, in some studies (including tsunami simulations), the vertical component of the velocity is ignored, and a different set of equations, called shallow-water equations, are solved for. Shallow-water equations are derived by integrating the velocity in the vertical direction, and can be written as

$$\frac{\partial \eta}{\partial t} + \frac{\partial \eta u}{\partial x} + \frac{\partial \eta v}{\partial x} = 0, \tag{8.138}$$

$$\frac{\partial(\eta u)}{\partial t} + \frac{\partial}{\partial x}(\eta u^2 + \frac{1}{2}g\eta^2) + \frac{\partial \eta uv}{\partial y} = \tau_x^{wind} - \tau_x^{Bottom}, \tag{8.139}$$

$$\frac{\partial(\eta v)}{\partial t} + \frac{\partial \eta uv}{\partial x} + \frac{\partial}{\partial y}(\eta v^2 + \frac{1}{2}g\eta^2) = \tau_y^{wind} - \tau_y^{Bottom}. \tag{8.140}$$

In Equations (8.138), (8.139), and (8.140), η is the surface elevation, u and v are the vertically integrated horizontal components of the velocity, τ^{wind} is the wind stress, and τ^{bottom} is the bottom friction, usually formulated as

$$\tau_x^{Bottom} = C_z^{-2} gu\sqrt{u^2 + v^2}, \tag{8.141}$$

$$\tau_y^{Bottom} = C_z^{-2} gv\sqrt{u^2 + v^2}, \tag{8.142}$$

where C_z is the Chezy number. If we solve the abovementioned equations simultaneously, we are able to obtain a 2-D solution for the flow.

8.5.4 Continuity Equation and Poisson Equation for Pressure

Consider that, in the case of three-dimensional models, Equation (8.127) provides three separate equations for use in deriving the velocity's different components (the pressure being an a priori known variable). The pressure can be calculated hydrostatically using Equation (8.138), which is derived via the application of the continuity equation. The majority of ocean models assume a hydrostatic pressure, and, in the case of coarse resolution modeling, the pressure is well-assumed to be hydrostatic. However, by increasing the resolution and going to finer and finer grid resolutions, the actual pressure deviates more than the purely hydrostatic pressure. This deviation is widely known as nonhydrostatic pressure, and, if the horizontal grid resolution of the model is of the order of just a few decimeters, then the nonhydrostatic portion of the pressure can no longer be ignored. There are many methods available for calculating nonhydrostatic pressure, and, although they all make use of the continuity equation in one way or another, the equation finally derived for the nonhydrostatic pressure is considered to be merely a mathematical equation, and is not considered to have any physical meaning.

The literature supplies a variety of formulations for nonhydrostatic pressure, with projection methods arguably being the most interesting class of these types of equations. Note that, for these equations, the pressure is calculated in a manner which ensures the velocity will be divergence-free. The fractional step method is a noteworthy example, as it evaluates all of the terms in the NS equation, except for the pressure effect, and also calculates an intermediate value for the velocity field. This intermediate velocity field does not necessarily satisfy the continuity equation: it may not be divergence-free; however, we can later use the discretized form of the equation to calculate the pressure in such a way that the velocity is guaranteed to be divergence-free for the next time step. This method requires the solution of a Poisson equation of the form

$$\nabla \cdot \nabla(p) = \frac{1}{\delta t}\nabla \cdot U_{intermediate}. \tag{8.143}$$

8.5.5 Numerical Solution of the Navier–Stokes Equation

Despite its importance and wide range of applications, mathematicians have not yet proven that solutions always exist in three dimensions, or that, if they do exist, that they contain any singularity. These questions are called the NS existence and smoothness problems, and the Clay Mathematics Institute offers a one million dollar prize to any person who can provide a solution or a counterexample.

Currently, the only viable means of solving for NS equations is via the utilization of one of the available numerical methods. One noteworthy example is the application of the mimetic method to the discretization of NS or shallow water equation as a means of obtaining a set of (non)linear equations. (Note that the quality of the discretization method being used affects both the accuracy of the solution and the computer program's performance.) One can solve the NS equation directly, via a direct numerical simulation (DNS) method. DNS methods are known to provide the most accurate solutions for the NS equation; however, they require a very fine grid, with the number of required grid cells proportionate to $Re^{9/4}$. Now, recalling that the Reynolds number is very high in geophysical flows, it is easy to conclude that the DNS approach is not feasible in this case. In fact, given current computer capacities and strengths, the DNS method is perforce limited to simple geometry and purely theoretical research.

The majority of ocean models currently use a method known as the Reynolds average numerical simulation (RANS); and, in this case, the velocity field and other variables are averaged in time. This leads to extra terms in the NS equations, and is known as the **turbulence model**. Another well-known approach is called **large eddy simulation** (LES); and, although it is more accurate than the RANS method, it is also more computationally demanding. In this approach, the velocity and other variables are filtered spatially, using a general filter kernel, and decomposed into both a mean part and a fluctuation part,

$$u = \bar{u} + u', \tag{8.144}$$

where $\bar{u}$ is the mean part and u' is the fluctuation part. By putting these decomposed variables back into the NS equation, we get

$$\frac{\partial \bar{u}_i}{\partial t} + \bar{u}_j \frac{\partial \bar{u}_i}{\partial x_j} + \frac{\partial \bar{p}}{\partial x_i} + \frac{\partial \tau_{ij}}{\partial x_j} - \frac{1}{Ro}\varepsilon_{ij3}\bar{u}_j + \frac{1}{Fr^2}\beta\delta_{i3} = 0, \tag{8.145}$$

where τ_{ij} is called the sub-grid scale (SGS) model. The SGS model takes into account the effect of the eddies that are smaller than the grid or filter size.

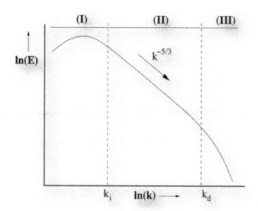

Figure 8.12: Energy spectrum of a three-dimensional flow.

Usually, simulation and modeling are used interchangeably, although they do not mean the same thing. In the LES approach, one has to distinguish between what is being simulated and what is being modeled. In LES, large eddies are simulated or resolved, while small eddies are modeled using a SGS model. This idea has its root in the energy spectrum of the flow (Figure 8.12). By looking at the energy spectrum, three different regions can be distinguished, as follows:

- Region I, Large-scale structures: This region is affected by the problem geometry, forcing, and conditions

- Region II, Inertial range: Always has the same slope, known as the 5/3 rule

- Region III, Viscous range: The energy is lost due to diffusion

This idea was prompted by the energy spectrum, wherein, if the flow's large structures are known, then its small structures can be reconstructed (modeled) based on those of the big structures, since they always follow the same slope. As a result, with the LES approach: large structures are simulated or resolved, and small structures are modeled; however, what is considered "large" or "small" here is a relative matter, and depends on the size of the filter chosen. In fact, LES accuracy and performance varies between those of DNS and RANS; such that, if you choose a very small filter size, LES will look more like DNS, whereas, if a big filter is chosen, then the LES method looks more like the RANS method (also called the **very large eddy simulation (VLES) method**). The three different approaches mentioned here, and the regions they simulate (compute), and model, are compared in Figure 8.5.5.

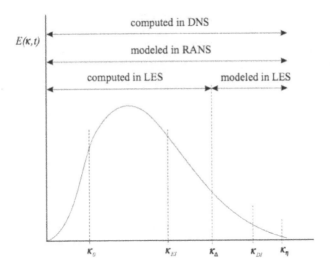

Figure 8.13: Simulated (computed) versus modeled region using different solution approaches.

8.5.6 The Mimetic Approach

As mentioned above, the only viable approach for solving NS equations is the utilization of numerical methods to discretize the governing equations into a system of (non)linear equations. For example: Mimetic operators provide all of the operators required for the discretization of the NS equation. At the Computational Science Research Center (CSRC), `http://www.csrc.sdsu.edu`, at San Diego State University (SDSU), a new class of model is currently under development, which is intended for use with geophysical flows. The model is called the **unified curvilinear ocean atmosphere model (UCOAM)**, and is intended for use in very high-resolution studies of oceanic and atmospheric flows. UCOAM makes use of a curvilinear grid in all three directions (including the vertical), and is designed for use in ultra-high-resolution studies; studies in which a horizontal resolution of only a few decameters is needed (making this model fully nonhydrostatic).

Recall that nonhydrostatic models end up with the solution of a Poisson equation for pressure. Solving the Poisson equation is time consuming; in fact, the current version of UCOAM spends most of its time solving this equation. Therefore, it is very important to use a discretization scheme that is efficient for solving this equation. Currently, the UCOAM model is using finite differences. However, efforts are underway to convert the model so that it harnesses Castillo–Grone's mimetic (CGM) difference operators.

CGM operators, particularly the gradient and divergence operators, have been shown to provide solutions with better accuracy. There have been many journal papers published on the performance of CGM operators in one-dimensional, regularly-spaced grids. However, there are few publications addressing the performance of CGM operators in two-dimensional, regularly-spaced grids, one-dimensional irregularly spaced grids, or in fully curvilinear two- and three-dimensional grids.

Consequently, efforts are currently underway to study the performance of CGM operators in both 2-D and 3-D curvilinear grids. In order to test the performance of the CGM operator, a fully curvilinear grid had to be generated (see Figure 8.14). The Poisson equation (8.146) was then solved using both a CGM difference operator and a conservative central difference discretization of the same equation.

$$\frac{\partial^2 f}{\partial x^2} + \frac{\partial^2 f}{\partial y^2} = \frac{\cos \sqrt{x^2 + y^2}}{\sqrt{x^2 + y^2}} - \sin \sqrt{x^2 + y^2}. \tag{8.146}$$

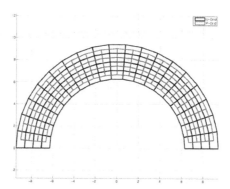

Figure 8.14: Two-dimensional curvilinear grid.

Equation 8.146 can be written using gradient and divergence operators as

$$\nabla \cdot (k \nabla f) = \frac{\cos \sqrt{x^2 + y^2}}{\sqrt{x^2 + y^2}} - \sin \sqrt{x^2 + y^2}, \tag{8.147}$$

where k, in the case of the Poisson equation and Cartesian grids, is equal to the identity matrix. However, one has to note, that, in the case of curvilinear grids, and due to the transformation of the equations from a curvilinear Cartesian grid to a computational grid, the abovementioned equation is transformed into

$$\tilde{\nabla} \cdot (\tilde{k} \tilde{\nabla} \tilde{f}) = \tilde{F}, \tag{8.148}$$

where ˜ means the variable or the operator in the computational domain. $\tilde{k}$ is no more equal to the identity matrix, and its entries are shown, as follows:

$$
\begin{aligned}
J\tilde{k}_{1,1} &= (y_\eta^2 + x_\eta^2) \\
J\tilde{k}_{2,2} &= (y_\xi^2 + x_\xi^2) \\
J\tilde{k}_{1,2} &= J\tilde{k}_{2,1} = -(y_\xi y_\eta + x_\xi x - \eta),
\end{aligned}
\tag{8.149}
$$

where J is the Jacobian of the transformation and $(x_\xi, x_\eta, y_\xi, y_\eta)$ are the metrics of the transformation. Figure 8.15 shows that the CGM operator produces better results.

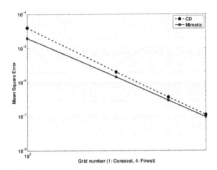

Figure 8.15: Mean square error comparison of the CGM difference operator (solid line) and the conservative central difference discretization (dashed line) of the Poisson equation.

The same behavior is seen in 3-D curvilinear coordinates. The right-hand side of Poisson's equation was changed to $e^{-\lambda(R-2\pi)/\pi}/(e^\lambda - 1)$, where $\lambda = -1$. A 3-D curvilinear grid was generated (see Figure 8.16), and the error was compared. As Figure 8.17 shows, once again, the CGM method performed better in 3-D.

8.5.7 Sample Problems

1. Using the buoyancy force, one can derive a measure for stability, as follows:

$$
E = \frac{1}{\rho}\frac{\partial \rho}{\partial z}.
\tag{8.150}
$$

One can also express the stability in terms of the stability frequency, known as the Brunt-Vaisala frequency, as follows:

$$
N^2 = -gE.
\tag{8.151}
$$

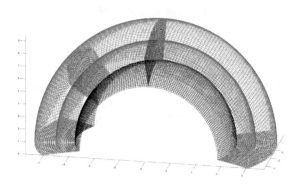

Figure 8.16: 3-D curvilinear grid.

(a) Discuss how E (and N^2) can be used to define stable, unstable, and neutral conditions? What value does E have in each region?

(b) Write a computer code that accepts a density profile and calculates the stability frequency.

2. In the nonhydrostatic calculation of pressure, one combines the continuity and the momentum equation to derive a Poisson's equation for the pressure. One example is

$$\nabla \cdot \nabla p = \frac{1}{\delta t} \nabla \cdot \mathbf{u}_{intermediate}. \qquad (8.152)$$

(a) Show how the equation above was derived.

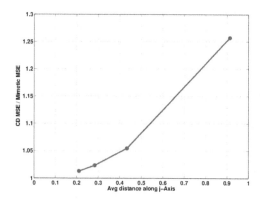

Figure 8.17: Mean square error comparison of CGM difference operator and conservative central difference discretization of the Poisson equation.

(b) Write a computer program that solves the equation above.

(c) Update the velocity field, as follows:

$$\frac{\partial u}{\partial t} = \nabla p, \tag{8.153}$$

and check if the velocity field is indeed divergence free after the update.

3. Consider a channel with constant fluid velocity ($\mathbf{u}$ is taken to be constant). Suppose, the temperature is suddenly reset at the end of the channel. The governing equation for temperature is

$$\frac{\partial T}{\partial t} + \mathbf{u}\frac{\partial T}{\partial x} = \alpha\frac{\partial^2 T}{\partial x^2}, \tag{8.154}$$

Assume a channel with 2.0 m length, $\alpha = 0.2$ m^2/s and $\mathbf{u} = u = 0.2$ m/s. Also, assume that the temperature is 20 °C at the beginning of the channel and 100 °C at the end.

(a) Write a computer program to solve the equation above for both the time and spatial components.

(b) Integrate the equation above using the computer code that you developed in Part 1 of this problem to derive the steady state solution.

(c) A steady solution can be obtained by solving the following steady state equation:

$$\mathbf{u}\frac{\partial T}{\partial x} = \alpha\frac{\partial^2 T}{\partial x^2}. \tag{8.155}$$

Obtain an analytic solution to the steady state equation presented above.

(d) Compare the steady state solution you obtained in Part 2 of this problem with the analytic solution obtained in Part 3 of this problem.

8.6 Concluding Remarks

In this chapter, we presented application problems in which mimetic discretization methods were successfully applied. We also presented a highly diverse collection of background knowledge.

We showed that the operators first proposed by Castillo and Grone (see [150]) ensured that the discretization of the boundary conditions is compatible (in regards to both the accuracy and the physics) with the interior computation, by avoiding any outer-domain or ghost points.

We also introduced important background knowledge for each considered application, in order for this to be a self-contained work. We never intended for each section to cover every specific detail, but we did intend for the sections to be as comprehensive and illustrative as possible, given the scope of our work.

Appendix A

Heuristic Deduction of the Extended Form of Gauss' Divergence Theorem

In this appendix, we present a heuristic approach with which to obtain the following formula:

$$\int_\sigma \rho \mathbf{v} \cdot \mathbf{n} \, d\sigma = \int_\Omega \nabla \cdot (\rho \mathbf{v}) \, dV. \tag{A.1}$$

The interested reader may also consult a formal proof in [185].

Now, let us consider a uniform steady fluid velocity field $v\mathbf{i}$, through a rectilinear horizontal channel with rectangular cross-section, originating a volume flux ϕ_V and a mass flux ϕ_M, crossing a planar surface S with area $A(S)$. (See Figure A.1.)

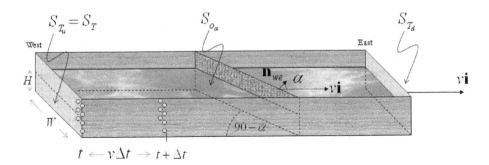

Figure A.1: Uniform liquid mass flow.

The flow direction defines a west-east sense, and both upstream (u) and downstream (d) transversal channel sections S_{T_u} and S_{T_d}, with bathed areas $A_{T_u} = WH$ and $A_{T_d} = A_{T_u}$. If we consider the liquid to be incompressible, then its mass density ρ is a constant.

Next, let us consider the volume flow and the mass flow through a transversal section S_{T_u}, which is denoted by $\phi_V(S_{T_u})$ and $\phi_M(S_{T_u})$, respectively. Since the fluid particles travel a distance $v\Delta t$ after crossing S_{T_u} at instant t, these concepts are defined, as follows:

$$\phi_V(S_{T_u}) \triangleq \frac{\Delta V}{\Delta t} = vA(S_{T_u}), \tag{A.2}$$

$$\phi_M(S_{T_u}) \triangleq \frac{\Delta M}{\Delta t} = \rho vA(S_{T_u}), \tag{A.3}$$

where ΔV and $\Delta M = \rho A(S_{T_u})v\Delta t$ are the volume and mass of the liquid that crosses section S_{T_u} during the interval $[t, t + \Delta t]$, and $\Delta V = A(S_{T_u})v\Delta t = WH(v\Delta t)$ (see Fig. A.1). The label u is suppressed from S_{T_u}, since the precise location of the upstream transversal section is arbitrary.

Now, if we consider an oblique section S_{o_α}, such that $\mathbf{n}_{we} \cdot \mathbf{i} = \cos\alpha$, where $\mathbf{n}_{we}$ is the unit normal to surface S_{o_α} in the west-east sense, we get:

$$\phi_V(S_{o_\alpha}) = v\cos\alpha\, A(S_{o_\alpha}) = v\mathbf{i} \cdot \mathbf{n}_{we}A(S_{o_\alpha}), \tag{A.4}$$

and

$$\phi_M(S_{o_\alpha}) = \rho\, v\cos\alpha\, A(S_{o_\alpha}) = \rho\, v\mathbf{i} \cdot \mathbf{n}_{we}A(S_{o_\alpha}). \tag{A.5}$$

Letting $\mathbf{v} = v\mathbf{i}$ and denoting $\phi_V(S_{o_\alpha}) = Q_{we}(S_{o_\alpha})$, we observe that, for a uniform flow ($\mathbf{v}$ constant along the channel), $Q_{we}(S_{o_\alpha}) = Q_{we}(S_T)$, since $A(S_{o_\alpha})\cos\alpha = A(S_T)$. This means that a uniform flow must correspond to the motion of an incompressible fluid, because if $Q_{we}(S_T) > Q_{we}(S_{o_\alpha})$, for example, then the fluid volume would accumulate per unit differential time.

It is important to note that

$$Q_{we}(S_{o_\alpha}) = A(S_{o_\alpha}) < \mathbf{v}, \mathbf{n}_{we} > . \tag{A.6}$$

Now, we can formulate the concept of volume flux Q_{we} (or volume flow per unit differential time traversing any **curved** and smooth surface σ, which is oriented from west to east (see Figure A.2.) Therefore, we assume that there is a uniquely defined continuous field $\mathbf{n}_{we}$ of normals throughout σ. This formulation will now be studied as a two-stage process of "analysis-synthesis."

For the Analysis Stage, we begin by **tessellating** the given curved surface σ, so that each surface portion $\Delta\sigma_i$ is small enough for the continuous vector field $\mathbf{v}$ to have a good constant approximation upon each $\Delta\sigma_i$. Then, taking $i = 1, 2, ..., n$, with a large enough n, $\Delta\sigma_i$ can be approximated by a tangent flat surface ΔF_i, with small area ΔA_i, and the whole surface

$$\sigma = \bigcup_{i=1}^{n} \Delta\sigma_i \tag{A.7}$$

is now approximated by a polyhedral surface F, composed of the small plane faces ΔF_i (see Figure A.3). Also, the velocity field $\mathbf{v}(\mathbf{x}, t)$ at some instant t is approximated by a field $\mathbf{v}^*(\Delta\sigma_i)$, which is **uniform** throughout ΔF_i, with $\mathbf{v}^*(\Delta\sigma_i) = \mathbf{v}(P_i, t)$, where P_i is the point of tangency of ΔF_i with $\Delta\sigma_i$ (see Figure A.4). Hence:

$$Q_{we}^*(\Delta F_i) = \Delta A_i < \mathbf{v}(P_i, t), \mathbf{n}_{we}(P_i) >, \tag{A.8}$$

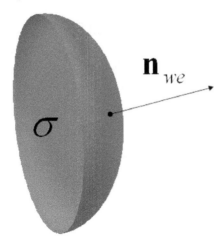

Figure A.2: Surface σ.

Figure A.3: Polyhedral surface F.

having designated the volume flux due to a field $\mathbf{v}^*$ by Q_{we}^*.

The approximation $Q_{we}(\Delta\sigma_i) \approx Q_{we}^*(\Delta F_i)$ is natural under the following two assumptions:

1. Continuity of the field $\mathbf{v}$ throughout 3-D space, which implies good constant approximation over small subdomains

2. Smoothness of the surface σ (differentiability at each point $P \in S$), implying good local plane approximations (see Figure A.4.)

For the <u>synthesis stage</u>, we now **sum** the contributions from the volume

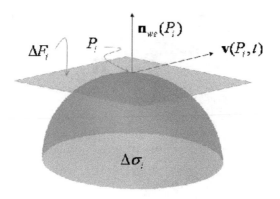

Figure A.4: Curved surface $\Delta\sigma_i$.

fluxes obtained by analysis in stage i); and, since we ask

$$Q_{we}(\sigma) = Q_{we}\left(\bigcup_{i=1}^{n} \Delta\sigma_i\right) = \sum_{i=1}^{n} Q_{we}(\Delta\sigma_i), \qquad (A.9)$$

we see that

$$\sum_{i=1}^{n} Q_{we}^*(\Delta F_i) \qquad (A.10)$$

approximates $Q_{we}(\sigma)$. Now, due to the form of each summand, namely,

$$Q_{we}^*(\Delta F_i) = \Delta A_i < \mathbf{v}(P_i, t), \mathbf{n}_{we}(P_i) >, \qquad (A.11)$$

we are reminded of **Riemann sums**, and, obviously ΔA_i approximates the area of the small surface element $\Delta\sigma_i$. These heuristic considerations finally lead us **to define** $Q_{we}(\sigma)$ as a surface integral:

$$Q_{we}(\sigma) = \int_{\sigma} < \mathbf{v}, \mathbf{n}_{we} > \, d\sigma. \qquad (A.12)$$

Analogously, we define $\dot{M}_{we}(\sigma)$, the **mass flow** through surface σ in the west-east direction, per unit differential time, as follows:

$$\dot{M}_{we}(\sigma) = \int_{\sigma} \rho < \mathbf{v}, \mathbf{n}_{we} > \, d\sigma. \qquad (A.13)$$

A.1   Standard Form of Gauss' Divergence Theorem

Recall that $\mathbf{v} = P\mathbf{i} + Q\mathbf{j} + R\mathbf{k}$. Now, consider a small rectangular parallelepiped with edge lengths Δx, Δy, and Δz. The net flux through the faces orthogonal to the x-axis is

$$< \mathbf{v}, \mathbf{i} > (x + \Delta x, y, z) - < \mathbf{v}, \mathbf{i} > (x, y, z) = \frac{\partial P}{\partial x}(x^*, y, z)\Delta y \Delta z \Delta x. \quad \text{(A.14)}$$

Analogous expressions for the other four faces yield as a total net flux of:

$$\left(\frac{\partial P}{\partial x}(x^*, y, z) + \frac{\partial Q}{\partial y}(x, y^*, z) + \frac{\partial R}{\partial z}(x, y, z^*) \right) \Delta x \Delta y \Delta z, \quad \text{(A.15)}$$

where $x < x^* < x + \Delta x$, $y < y^* < y + \Delta y$ and $z < z^* < z + \Delta z$. The limiting sum of these net fluxes over all 3-D cells filling a given 3-D region Ω analogously define a Riemann integral:

$$\int_{\Omega} \left(\frac{\partial P}{\partial x} + \frac{\partial Q}{\partial y} + \frac{\partial R}{\partial z} \right) dV = \int_{\Omega} \nabla \cdot (\mathbf{v}) \, dV. \quad \text{(A.16)}$$

Since there is no creation or destruction of fluid flux at the interior of Ω, this last quantity equals the total net flux through the enclosing boundary surface σ, therefore:

$$\int_{\sigma} \mathbf{v} \cdot \mathbf{n} \, d\sigma = \int_{\Omega} \nabla \cdot (\mathbf{v}) \, dV. \quad \text{(A.17)}$$

Notice that, in the case of a region Ω totally enclosed by an orientable surface σ, there is a natural orientation for $\mathbf{n}$, namely, the outward one. Therefore, we consider $\mathbf{n}_{we} = \mathbf{n}$.

This formula, under adequate smoothness requirements for σ, and $\mathbf{v}$ is known as **Gauss' divergence theorem**.

A.2   Extended Form of Gauss' Divergence Theorem

Consider the auxiliary field $\mathbf{w} = \rho \mathbf{v}$, so that $< \mathbf{w}, \mathbf{n} >= \rho < \mathbf{v}, \mathbf{n} >$, with ρ a given scalar field. When $\nabla \cdot (\rho \mathbf{v})$ is expanded as the sum of two terms, namely, $< \nabla \rho, \mathbf{v} >$ and $\rho \nabla \cdot \mathbf{v}$ (one of which exhibits a gradient), then the outward surface mass flow $M(\sigma)$ equals the sum of two triple integrals:

$$\int_{\sigma} \rho < \mathbf{v}, \mathbf{n} > d\sigma = \int_{\Omega} < \nabla \rho, \mathbf{v} > dV + \int_{\Omega} \rho \nabla \cdot (\mathbf{v}) dV. \quad \text{(A.18)}$$

This formula, obtained as a corollary to the previous one, assuming the scalar field ρ to be continuously differentiable throughout Ω, will be referred to as the **Extended form of Gauss' divergence theorem**.

Of course, the field involved in the evaluation of the triple integrals must behave in such a way at the boundary $\sigma = \partial \Omega$ that it assures convergence of the possibly improper Riemann volume integrals. Continuity of $\nabla\rho$ and $\nabla\cdot\mathbf{v}$ on σ is certainly sufficient, as is the case in many applications, but is by no means necessary.

Appendix B

Tensor Concept: An Intuitive Approach

The word "tensor" has a physically motivated root: namely the word "tension". In Continuum mechanics, it is customary to classify the forces acting upon differential volume elements dV. One class corresponds to **body forces** $\mathbf{b}$, defined as external forces per unit differential mass $dM = \rho dV$, where ρ is the mass density of the continuous media, so that the corresponding external forces per unit differential volume is $\rho\mathbf{b}$. Typical examples are gravitational forces, with $\mathbf{b} = -g\mathbf{k}$, and centrifugal forces, due to rotation.

The other class corresponds to the **surface forces** acting upon the surface bounding dV.

In order to isolate the differential volume element dV from the rest of the continuous media, it is necessary to consider the bounding surface as a mathematically ideal surface. This gives rise to two surfaces, which are congruent to the bounding surface. One of them can be thought of as made of molecules within the volume element dV, and the other can be considered as made of molecules lying inside the set-complement. These two sets of molecules act upon each other, thus exerting force throughout the separating surface, according to Newton's Law of Action and Reaction. Therefore, in the case of an elastic medium being stretched, there will be one force per unit differential surface area $d\sigma$, $(\mathbf{t}d\sigma)$ as the **traction force** exerted from the complement set to dV, and a **contraction force** $-\mathbf{t}d\sigma$ exerted by the boundary molecules from dV toward the complement set. Such vector field $\mathbf{t}$ is generally designated as the **tension**.

In cases involving nonviscous fluids, the surface forces do not act as tractions. As it is familiar, the liquid pressure forces act upon surfaces (such as a diver's eardrum) in a **compressive way**, and, in a direction normal to the surface, so that in case of $\mathbf{t}d\sigma = -(p\mathbf{n})d\sigma$, where $\mathbf{n}$ is a unit normal outwardly directed with respect to dV.

From this simple definition, it might appear that the scalar-valued pressure field p might depend not only on the space location for $d\sigma$, but also on the direction of the normal $\mathbf{n}$ to dV.

Pascal's principle states that the pressure at a point within an inviscid fluid, i.e., a liquid free from tangential surfaces, **is the same in all directions**. We will prove this in detail, since the geometry and symbols needed to define the general Cartesian tensor are already present in this very simple and familiar situation.

Suppose that the pressure force $p\mathbf{n} = p_n\mathbf{n}$ is different for $\mathbf{n} = \mathbf{i}$, $\mathbf{j}$, $\mathbf{k}$, so that we are dealing with four scalar values p_n, p_x, p_y, p_z at the same point $p = (x, y, z)$ (See Figure B.1).

Now consider the differential volume dV for a tetrahedron $PABC$, where $A = (x + \Delta x, y, z)$, $B = (x, y + \Delta y, z)$, and $C = (x, y, z + \Delta z)$.

Let $d\sigma_x$, $d\sigma_y$, $d\sigma_z$, and $d\sigma_n$ denote the areas of the boundary triangular faces PBC, PCA, PAB, and ABC, respectively (see Figure B.1.)

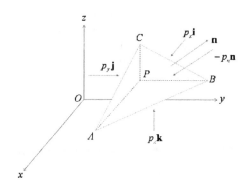

Figure B.1: Four scalar values for pressure acting at the same point $p = (x, y, z)$.

The triangular faces of the tetrahedron $PABC$, which are parallel to the coordinate planes, can be regarded as projections of the face ABC with normal $\mathbf{n}$ outwardly directed from the interior of $PABC$. Therefore,

$$d\sigma_x = d\sigma_n < \mathbf{n}, \mathbf{i} >, \tag{B.1}$$
$$d\sigma_y = d\sigma_n < \mathbf{n}, \mathbf{j} >, \tag{B.2}$$
$$d\sigma_z = d\sigma_n < \mathbf{n}, \mathbf{k} > . \tag{B.3}$$

The volume of $PABC$ can be expressed as:

$$dV = \frac{1}{3}d\sigma_x(\Delta x) = \frac{1}{3}d\sigma_y(\Delta y) = \frac{1}{3}d\sigma_z(\Delta z) = \frac{1}{6}\Delta x \Delta y \Delta z. \tag{B.4}$$

Newton's law for momentum conservation applied to the liquid contained in $PABC$ under the surface forces $-p_n d\sigma_n \mathbf{n}$, $p_x d\sigma_x \mathbf{i}$, $p_y d\sigma_y \mathbf{j}$ and $p_z d\sigma_z \mathbf{k}$, yields

$$p_x d\sigma_x \mathbf{i} + p_y d\sigma_y \mathbf{j} + p_z d\sigma_z \mathbf{k} - p_n d\sigma_n \mathbf{n} + \rho^* dV \mathbf{b} = \rho^* dV \mathbf{a}. \tag{B.5}$$

Here, ρ^* is the average mass-density of the liquid contained in $PABC$; p_x, p_y, p_z, and p_n stand for the average pressure over the corresponding triangular surfaces, and $\mathbf{a}$ is the acceleration of its center of gravity.

If we use the expressions for the projected triangular areas, then the previous equation now yields

$$d\sigma_n \left(p_x < \mathbf{n}, \mathbf{i} > \mathbf{i} + p_y < \mathbf{n}, \mathbf{j} > \mathbf{j} + p_z < \mathbf{n}, \mathbf{k} > \mathbf{k} - p_n \mathbf{n} \right) = \rho^* dV (\mathbf{a} - \mathbf{b}).$$
(B.6)

If we let the diameter of $PABC$ tend to zero, keep $\mathbf{n}$ constant, and observe that $dV/d\sigma_n$ also tends to zero, we conclude that

$$p_x < \mathbf{n}, \mathbf{i} > \mathbf{i} + p_y < \mathbf{n}, \mathbf{j} > \mathbf{j} + p_z < \mathbf{n}, \mathbf{k} > \mathbf{k} - p_n \mathbf{n} = 0\mathbf{i} + 0\mathbf{j} + 0\mathbf{k} = \mathbf{0}. \quad \text{(B.7)}$$

By taking the scalar products with $\mathbf{i}$, $\mathbf{j}$, and $\mathbf{k}$ from this vector equation, we conclude that $p_x = p_y = p_z = p_n$ at point $p = (x, y, z)$, which is just Pascal's principle.

B.1 Newton's Law Applied to the Mass in the Tetrahedron

When considering viscous fluids or elastic solids, tangential components appear in the surface forces so that we must now consider, total surface forces per unit area $\mathbf{t}_n$, $(-\mathbf{t}_x)$, $(-\mathbf{t}_y)$, and $(-\mathbf{t}_z)$, with:

$$\mathbf{t}_x = \tau_{xx}\mathbf{i} + \tau_{xy}\mathbf{j} + \tau_{xz}\mathbf{k}, \quad \text{(B.8)}$$
$$\mathbf{t}_y = \tau_{yx}\mathbf{i} + \tau_{yy}\mathbf{j} + \tau_{yz}\mathbf{k}, \quad \text{(B.9)}$$
$$\mathbf{t}_z = \tau_{zx}\mathbf{i} + \tau_{zy}\mathbf{j} + \tau_{zz}\mathbf{k}, \quad \text{(B.10)}$$
$$\mathbf{t}_n = \tau_{nx}\mathbf{i} + \tau_{ny}\mathbf{j} + \tau_{nz}\mathbf{k}. \quad \text{(B.11)}$$

See Figure B.2. Note that $\mathbf{t}_n$ now appears as **outwardly** directed from the interior of $PABC$, being thought of as a **traction** instead of as a compression, as in the simple pressure case. Now Newton's law yields

$$-\mathbf{t}_x d\sigma_x - \mathbf{t}_y d\sigma_y - \mathbf{t}_z d\sigma_z + \mathbf{t}_n d\sigma_n = \rho^* dV (\mathbf{a} - \mathbf{b}). \quad \text{(B.12)}$$

Substituting $d\sigma_x = d\sigma_n < \mathbf{n}, \mathbf{i} >$ and so on, we get

$$d\sigma_n \left(-\mathbf{t}_x < \mathbf{n}, \mathbf{i} > -\mathbf{t}_y < \mathbf{n}, \mathbf{j} > -\mathbf{t}_z < \mathbf{n}, \mathbf{k} > +\mathbf{t}_n \right) = \rho^* dV (\mathbf{a} - \mathbf{b}). \quad \text{(B.13)}$$

Since:

$$\lim_{\text{Diam } PABC \to 0} \frac{dV}{d\sigma_n} = 0, \quad \text{(B.14)}$$

then

$$\mathbf{t}_n = \mathbf{t}_x < \mathbf{n}, \mathbf{i} > +\mathbf{t}_y < \mathbf{n}, \mathbf{j} > +\mathbf{t}_z < \mathbf{n}, \mathbf{k} > . \quad \text{(B.15)}$$

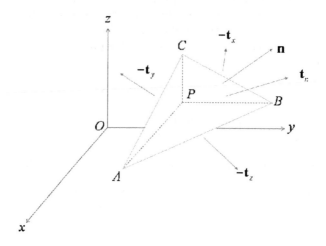

Figure B.2: Nine scalar values for a traction force acting at the same point $P = (x, y, z)$.

Taking the scalar product of this vector equation with $\mathbf{i}$, $\mathbf{j}$, and $\mathbf{k}$:

$$< \mathbf{t}_n, \mathbf{i} > = \tau_{xx} < \mathbf{n}, \mathbf{i} > +\tau_{yx} < \mathbf{n}, \mathbf{j} > +\tau_{zx} < \mathbf{n}, \mathbf{k} >, \qquad \text{(B.16)}$$

$$< \mathbf{t}_n, \mathbf{j} > = \tau_{xy} < \mathbf{n}, \mathbf{i} > +\tau_{yy} < \mathbf{n}, \mathbf{j} > +\tau_{zy} < \mathbf{n}, \mathbf{k} >, \qquad \text{(B.17)}$$

$$< \mathbf{t}_n, \mathbf{k} > = \tau_{xz} < \mathbf{n}, \mathbf{i} > +\tau_{yz} < \mathbf{n}, \mathbf{j} > +\tau_{zz} < \mathbf{n}, \mathbf{k} > . \qquad \text{(B.18)}$$

These three scalar equations can be re-written in matrix form:

$$\mathbf{t}_n = \tau \mathbf{n}, \qquad \text{(B.19)}$$

where $\mathbf{n}$ and $\mathbf{t}_n$ are 3-by-1 matrices, and τ is a 3-by-3 matrix:

$$\mathbf{n} = \begin{bmatrix} < \mathbf{n}, \mathbf{i} > \\ < \mathbf{n}, \mathbf{j} > \\ < \mathbf{n}, \mathbf{k} > \end{bmatrix} \quad \mathbf{t}_n = \begin{bmatrix} < \mathbf{t}_n, \mathbf{i} > \\ < \mathbf{t}_n, \mathbf{j} > \\ < \mathbf{t}_n, \mathbf{k} > \end{bmatrix} \quad \tau = \begin{bmatrix} \tau_{xx} & \tau_{yx} & \tau_{zx} \\ \tau_{xy} & \tau_{yy} & \tau_{zy} \\ \tau_{xz} & \tau_{yz} & \tau_{zz} \end{bmatrix} . \qquad \text{(B.20)}$$

The matrix τ is usually denoted as the Cartesian **Stress tensor**.

Therefore, in the particular case of an inviscid fluid, our matrix τ becomes $DIAG\,(-p, -p, -p)$.

At a given point $P = (x, y, z)$ **within** the continuous media being studied, one might consider infinitely many planes while varying the normal direction $\mathbf{n}$, but the knowledge of only the **nine** components of the stress tensor τ at P suffices for evaluating the surface force per unit area $\mathbf{t}_n = \tau \mathbf{n}$. Furthermore, under normal circumstances, one can prove that τ is a symmetric matrix, so that **only six** components of τ at P are needed to characterize the state of stress.

Also note that the usual matrix multiplication induces the **tensor product** of the tensor τ times a vector $\mathbf{n}$ to yield a vector $\mathbf{t}_n$.

Using the familiar indexing for matrices, if $\tau = \{\tau_{pq}\}$, $p, q = 1, 2, 3$ and $\mathbf{n} = \{n_q\}$, $\mathbf{t}_n = \{t_p\}$, then

$$t_p = \sum_{q=1}^{3} \tau_{pq} n_q. \tag{B.21}$$

The symmetry of the stress tensor τ is now expressed as $\tau_{pq} = \tau_{qp}$.

Tensors also appear in many applied fields, such as heat transfer and porous flows in oil reservoir modeling (see Chapter 8) and Einstein's gravitational gheory and many others.

In these physical situations, there is some scalar field f, such as temperature T or pressure p, and *grad* f is the basic flux vector density, which acts differently in different spatial directions. For instance, if μ is the viscosity of the fluid in a porous medium, then the pressure gradient ∇p can be related to the flow rate or Darcy velocity $\mathbf{v}$ by means of Darcy's law, as

$$\mathbf{v} = \left(-\frac{1}{\mu}\right) k \nabla p, \tag{B.22}$$

where k is the **permeability tensor of the rock**, which takes the diagonal form:

$$k = \begin{bmatrix} k_x & 0 & 0 \\ 0 & k_y & 0 \\ 0 & 0 & k_z \end{bmatrix}. \tag{B.23}$$

B.2   Newton's Law on a Parallelepiped

Note that we can also consider the **divergence of a tensor field**, as motivated by the application of Newton's law to the continuous media within an infinitesimal rectangular parallelepiped, yielding:

$$\rho \mathbf{a} = \nabla \cdot \tau + \rho \mathbf{b}. \tag{B.24}$$

This vector equation can be deduced, as follows: consider a far vertex $Q = (x + \Delta x, y + \Delta y, z + \Delta z)$ besides the nearest vertex $P = (x, y, z)$, as in Figure B.3. We deduce the k-component of the mentioned vector equation. Considering all the surface forces acting on the six faces of the parallelepiped,

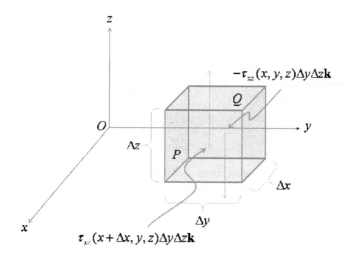

Figure B.3: Momentum equation to induce the divergence of a tensor field.

together with the body forces along the **k**-direction:

$$[\tau_{xz}(x + \Delta x, y, z) - \tau_{xz}(x, y, z)]\Delta y \Delta z$$
$$+[\tau_{yz}(x, y + \Delta y, z) - \tau_{yz}(x, y, z)]\Delta x \Delta z$$
$$+[\tau_{zz}(x, y, z + \Delta z) - \tau_{zz}(x, y, z)]\Delta x \Delta y =$$
$$\rho \Delta x \Delta y \Delta z(< \mathbf{a} - \mathbf{b}, \mathbf{k} >). \qquad (B.25)$$

Dividing by $\Delta x \Delta y \Delta z$ and taking the limit as $max(\Delta x, \Delta y, \Delta z)$ tends to zero,

$$\frac{\partial \tau_{xz}}{\partial x} + \frac{\partial \tau_{yz}}{\partial y} + \frac{\partial \tau_{zz}}{\partial z} = \rho < \mathbf{a}, \mathbf{k} > -\rho < \mathbf{b}, \mathbf{k} > . \qquad (B.26)$$

Recalling the symmetry of the three-by-three matrix τ, the last equation can be rewritten as:

$$\frac{\partial \tau_{zx}}{\partial x} + \frac{\partial \tau_{zy}}{\partial y} + \frac{\partial \tau_{zz}}{\partial z} = \rho < \mathbf{a} - \mathbf{b}, \mathbf{k} > . \qquad (B.27)$$

Analogously:

$$\frac{\partial \tau_{yx}}{\partial x} + \frac{\partial \tau_{yy}}{\partial y} + \frac{\partial \tau_{yz}}{\partial z} = \rho < \mathbf{a} - \mathbf{b}, \mathbf{j} > . \qquad (B.28)$$

$$\frac{\partial \tau_{xx}}{\partial x} + \frac{\partial \tau_{xy}}{\partial y} + \frac{\partial \tau_{xz}}{\partial z} = \rho < \mathbf{a} - \mathbf{b}, \mathbf{i} > . \qquad (B.29)$$

Therefore, **we can define** the three-by-one matrix $\nabla \cdot \tau$, as follows:

$$\nabla \cdot \tau = \begin{bmatrix} \nabla \cdot \mathbf{t}_x \\ \nabla \cdot \mathbf{t}_y \\ \nabla \cdot \mathbf{t}_z \end{bmatrix}, \tag{B.30}$$

where

$$\tau = \begin{bmatrix} \mathbf{t}_x \\ \mathbf{t}_y \\ \mathbf{t}_z \end{bmatrix}, \tag{B.31}$$

and $\mathbf{t}_x$, $\mathbf{t}_y$, and $\mathbf{t}_z$ are one-by-three matrices:

$$\mathbf{t}_x = [\tau_{xx} \ \tau_{xy} \ \tau_{xz}], \tag{B.32}$$

$$\mathbf{t}_y = [\tau_{yx} \ \tau_{yy} \ \tau_{yz}], \tag{B.33}$$

$$\mathbf{t}_z = [\tau_{zx} \ \tau_{zy} \ \tau_{zz}]. \tag{B.34}$$

Here we have considered only doubly indexed tensor components or, **second-order tensors** τ_{pq} as they can be also referred to, together with Cartesian coordinates. Higher-order tensors, regarded as fields defined in general curvilinear coordinates can also be defined, as the reader may easily find in the literature.

Appendix C

Total Force Due to Pressure Gradients

Consider the following three vector fields:

$$\mathbf{w}_1 = p\mathbf{i}, \tag{C.1}$$

$$\mathbf{w}_2 = p\mathbf{j}, \tag{C.2}$$

$$\mathbf{w}_3 = p\mathbf{k}. \tag{C.3}$$

Next, apply Gauss' formula:

$$\int_\Omega div\ \mathbf{w}_j dV = \int_\sigma \mathbf{w}_j \cdot (\mathbf{n}d\sigma) \text{ with } j = 1, 2, 3, \tag{C.4}$$

to obtain

$$\int_\Omega \frac{\partial p}{\partial x} dV = \int_\sigma p < \mathbf{i}, \mathbf{n} > d\sigma, \tag{C.5}$$

$$\int_\Omega \frac{\partial p}{\partial y} dV = \int_\sigma p < \mathbf{j}, \mathbf{n} > d\sigma, \tag{C.6}$$

$$\int_\Omega \frac{\partial p}{\partial z} dV = \int_\sigma p < \mathbf{k}, \mathbf{n} > d\sigma. \tag{C.7}$$

If we consider the left-hand sides of these equalities as Cartesian components of a vector, we obtain

$$\int_\Omega grad\ p\ dV = \int_\sigma p(\mathbf{n})\ d\sigma, \tag{C.8}$$

since

$$\mathbf{n} = < \mathbf{n}, \mathbf{i} > \mathbf{i} + < \mathbf{n}, \mathbf{j} > \mathbf{j} + < \mathbf{n}, \mathbf{k} > \mathbf{k}. \tag{C.9}$$

As a matter of general interest, note that we can immediately obtain Archimedes' buoyancy principle for a body totally immersed in a liquid with constant density ρ, in which case, $grad\ p = -\rho g$, so that the buoyancy force reads

$$\mathbf{F}(\sigma) = - \int_\sigma p(\mathbf{n}d\sigma) = \int_\Omega \rho g\ dV = g \int_\Omega \rho dV = gM = W, \tag{C.10}$$

where W is the weight of the displaced mass of fluid.

Appendix D

Heuristic Deduction of Stokes' Formula

Consider a two-dimensional vector field $\mathbf{v} = P\mathbf{i} + Q\mathbf{j}$, so that

$$\text{curl } \mathbf{v} = \left(\frac{\partial Q}{\partial x} - \frac{\partial P}{\partial y} \right) \mathbf{k}. \tag{D.1}$$

Let C be a closed, smooth curve bounding a planar surface σ, placed on the x-y plane. After orienting C counter clockwise, the unit normal $\mathbf{n}$ to σ points in the same direction as $\mathbf{k}$, i.e., $\mathbf{n} = \mathbf{k}$, and

$$\int_\sigma \mathbf{n} \cdot \text{curl } \mathbf{v} d\sigma = \int_\sigma \mathbf{k} \cdot \left(\frac{\partial Q}{\partial x} - \frac{\partial P}{\partial y} \right) \mathbf{k} \, d\sigma = \int_\sigma \left(\frac{\partial Q}{\partial x} - \frac{\partial P}{\partial y} \right) \, d\sigma. \tag{D.2}$$

Consider a cylindrical body Ω of unit height and cross-section σ, and an auxiliary three-dimensional vector field $\mathbf{w} = Q\mathbf{i} - P\mathbf{j} + 0\mathbf{k}$. (See Figure D.1.)

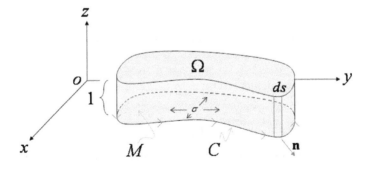

Figure D.1: An arbitrary clockwise oriented curve C and its related cylindrical body Ω.

Gauss' theorem applied to $\mathbf{w}$ in Ω yields

$$\int_\Omega div\ \mathbf{w}\ dV = \int_\Omega \left(\frac{\partial Q}{\partial x} - \frac{\partial P}{\partial y}\right) dxdydz = \tag{D.3}$$

$$\int_\sigma \left(\frac{\partial Q}{\partial x} - \frac{\partial P}{\partial y}\right) d\sigma = \int_M <\mathbf{w}, \mathbf{n}> dS. \tag{D.4}$$

where dS is the surface area element of the cylindrical mantle M. (See Figure D.1.) Since there is no depending on $\mathbf{w}$ and $\mathbf{n}$ from the vertical z-coordinate, then $dS = 1ds$, where ds is the arc length element along C, so that

$$\int_M <\mathbf{w}, \mathbf{n}> dS = \int_C <Q\mathbf{i} - P\mathbf{j}, n_x\mathbf{i} + n_y\mathbf{j}> ds \tag{D.5}$$

$$= \int_C (Qn_x - Pn_y)ds. \tag{D.6}$$

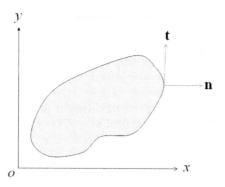

Figure D.2: Unit tangent vector $\mathbf{t}$ to curve C.

However, the unit tangent vector $\mathbf{t}$ to C can be written as $\mathbf{t} = -n_y\mathbf{i} + n_x\mathbf{j}$ (see Fig. D.2), so that:

$$Qn_x - Pn_y =< P\mathbf{i} + Q\mathbf{j}, -n_y\mathbf{i} + n_x\mathbf{j} >=< \mathbf{v}, \mathbf{t} > . \tag{D.7}$$

Therefore, we have obtained the following particular case of Stokes' formula for $\mathbf{v}$ and σ:

$$\int_\sigma \mathbf{n} \cdot curl\ \mathbf{v}\ d\sigma = \int_C <\mathbf{v}, \mathbf{t}> ds = \int_C \mathbf{v} \cdot d\mathbf{r}, \tag{D.8}$$

where $d\mathbf{r} = \mathbf{t}ds$.

Next, we consider a general differentiable 3-D vector field $\mathbf{v}$ and polyhedral tesselated surface $F = \bigcup_i F_i$, where each F_i, $i = 1, 2, ..., n$ is a flat surface element with a closed boundary curve C_i, so that

$$\int_{F_i} \mathbf{n} \cdot curl\ \mathbf{v}\ d\sigma = \int_{C_i} \mathbf{v} \cdot d\mathbf{r}. \tag{D.9}$$

Adding all these equalities yields

$$\int_{F} \mathbf{n} \cdot curl\ \mathbf{v}\ d\sigma = \int_{C} \mathbf{v} \cdot d\mathbf{r}, \tag{D.10}$$

where C is the boundary line supporting F, since the inner portions of C_i are traversed in opposite directions when considered as bounding adjacent surface elements (see Fig. D.3).

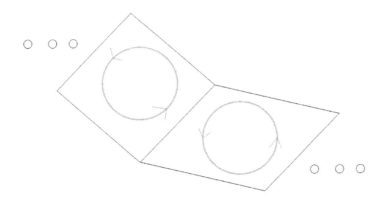

Figure D.3: Collection of curves C_i, $i = 1, 2, ..., n$.

A general, smooth surface σ can be approximated by a sequence of polyhedral surfaces with an increasing number n of flat surface elements, so that, in the limit:

$$\int_{\sigma} \mathbf{n} \cdot curl\ \mathbf{v}\ d\sigma = \int_{C} \mathbf{v} \cdot d\mathbf{r}. \tag{D.11}$$

Appendix E

Curl in a Rotating Incompressible Inviscid Liquid

A good intuitive grasp of the physical meaning of *curl* **v** can be attained by studying the velocity field of some incompressible inviscid liquid mass partially filling a vertical and cylindrical container with radius a, with its axis coincident with the z-coordinate axis, its bottom circular base lying in the plane $z = -H$, and its total height being $(H + h)$ (see Figure E.1). The liquid, initially at rest, fills the container up to the plane $z = 0$. Then, the liquid massively rotates when the cylindrical container is forced to rotate around its axis with a constant angular velocity w.

The liquid free surface deviates from the original static shape $z = 0$, and looks like a paraboloid of revolution after attaining a steady state. It is be assumed that the points at the liquid free surface experience atmospheric pressure throughout, because the height variations of the surface are relatively small when compared with the atmospheric height. This constant reference pressure can be taken to be zero.

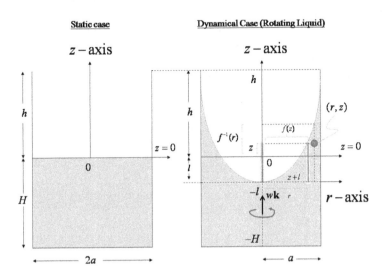

Figure E.1: Curl in a rotating incompressible inviscid liquid.

Let $z = -l$ denote the lowest point on the free surface of the rotating liquid, with $l = l(w)$, and let $z = h$ denote the height of the highest points ($h = h(w)$). Now, consider cylindrical coordinates (r, θ, z). Let $r = f(z)$ describe the shape of the rotating free surface for $-l \leq z \leq h$, so that $f(-l) = 0$ and $f(h) = a$. For each $z \in [-l, h]$, there is liquid for $f(z) \leq r \leq a$, and the liquid pressure $p = p(r, z)$ in the domain, since there is no θ-dependence. Clearly, $\mathbf{v} = \omega\hat{\mathbf{k}} \times (x\mathbf{i} + y\mathbf{j})$, so that $curl\ \mathbf{v} = 2\omega\hat{\mathbf{k}}$ is in the direction of the rotational axis, and its magnitude norm equals twice the angular velocity ω throughout the liquid mass.

The liquid density is some constant ρ, due to the incompressibility assumption. Also, the conservation of momentum applied to a rotating differential element of a liquid yields $k\rho(d\mathbf{v}/dt) = -\rho g\hat{\mathbf{k}} - \nabla p(r, z) = -(\partial p/\partial z)\hat{\mathbf{r}} - (\partial p/\partial z)\hat{\mathbf{k}} - \rho g\hat{\mathbf{k}}$, while $\mathbf{v}(x, y) = \omega\hat{\mathbf{k}} \times (x\mathbf{i} + y\mathbf{j})$ implies $d\mathbf{v}/dt = -\omega^2(x\mathbf{i} + y\mathbf{j}) = -\omega^2 r\hat{\mathbf{r}}$, so that $-\rho\omega^2 r\hat{\mathbf{r}} = -(\partial p/\partial z)\hat{\mathbf{r}} - (\partial p/\partial z)\hat{\mathbf{k}} - \rho g\hat{\mathbf{k}}$; yielding two differential equations:

$$\frac{\partial p}{\partial r} = \rho\omega^2 r, \ f(z) \leq r \leq a, \tag{E.1}$$

$$\frac{\partial p}{\partial z} = -\rho g; \tag{E.2}$$

for fixed $r \in [f(z), a]$, $z \in [-H, f^{-1}(r)]$, and $p(r, z) = -\rho g z + \phi(r)$. But $p(r, f^{-1}(r)) = 0$, so that $0 = -\rho g f^{-1}(r) + \phi(r)$ or $p(r, z) = -\rho g z + \rho g f^{-1}(r)$.

On the other hand, for fixed z, $p(r, z) = (1/2)(\rho\omega^2 r^2) + \Psi(z)$. Besides, $f^{-1}(0) = -l$ implies that $p(0, z) = -\rho g z - \rho g l$, and, therefore $\Psi(z) = p(0, z) = -\rho g z - \rho g l$ and $p(r, z) = (1/2)(\rho\omega^2 r^2) - \rho g z - \rho g l$. Finally, along the curve $r - f(z) = 0$, we must have $p = 0$, and the equation of the free surface is: $z + l = (\omega^2/2g)(x^2 + y^2)$ and $h + l = (\omega^2/2g)a^2$; therefore, knowing a and ω allows us to obtain $(h + l)(\omega)$. One can separate the values h and l using the fact that the volume under the paraboloid must equal $\pi a^2 H$, since the liquid is incompressible.

With numerical methods, real-life free surfaces can be approximated in shapes with high accuracy, such as ocean waves hitting ships and bouncing back.

Note that one should not believe, based on this example, that streamlines must be curved in order to have a nonzero curl.

Appendix F

Curl in Poiseuille's Flow

A simple example of a velocity field **v** having a nonzero curl, in spite of the fact that its streamlines define a family of parallel straight lines, is provided by **Poiseuille's flow**, which is a laminar (nonturbulent), steady, viscous flow at low velocities in a cylindrical tube, and is subject to a given constant pressure gradient along its axis (see Figure F.1.)

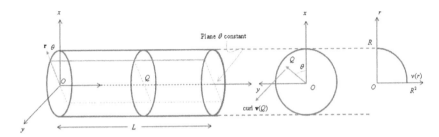

Figure F.1: Viscous flow at low velocities in a cylindrical tube.

The velocity field **v** can be exactly obtained by using cylindrical coordinates, with the positive z-axis coincident with the cylinder axis and also in the flow direction, and $dp/dz = $ constant $C < 0$. The field **v** does not depend upon z or θ, so that $\mathbf{v}(r) = v(r)\hat{\mathbf{k}}$. It is assumed that $v(R) = 0$ ("no slip" BC), and

$$\frac{dv}{dR}(0) = 0, \tag{F.1}$$

which models the maximum velocity at the tube's center.

Now, if we take a length L along the tube in order to measure the negative pressure gradient $[p(L) - p(0)]/L = C$, then

$$\frac{C}{\eta} = \frac{1}{r}\frac{d}{dr}\left(r\frac{dv}{dr}\right). \tag{F.2}$$

and the velocity field inside the cylinder of radius R is

$$\mathbf{v}(r) = -\frac{1}{4\eta} \cdot C \cdot (R^2 - r^2)\hat{\mathbf{k}}, \tag{F.3}$$

with $r^2 = x^2 + y^2$, where η is the **dynamic fluid viscosity**. For simplicity's sake, assume that both the pressure gradient and η have numerical values, such that $\mathbf{v}(r) = (R^2 - x^2 - y^2)\hat{\mathbf{k}}$, with $(curl\ \mathbf{v})(x, y) \cdot \mathbf{i} = -2y$, $(curl\ \mathbf{v})(x, y) \cdot \mathbf{j} = 2x$, and $(curl\ \mathbf{v})(x, y) \cdot \mathbf{k} = 0$; hence, $(curl\ \mathbf{v})(x, y) = curl\ \mathbf{v}(x\mathbf{i} + y\mathbf{j}) = -2y\mathbf{i} + 2x\mathbf{j} = 2r\hat{\mathbf{k}} \times \hat{\mathbf{r}}$, with $\hat{\mathbf{r}} = (x\mathbf{i} + y\mathbf{j})/r = \cos\theta\mathbf{i} + \sin\theta\mathbf{j}$, so we see that $curl\ \mathbf{v}(0, 0) = 0$. However, $curl\ \mathbf{v}(x\mathbf{i} + y\mathbf{j}) = 2\hat{\mathbf{k}} \times (x\mathbf{i} + y\mathbf{j})$, so that, if $\overrightarrow{OQ} = x\mathbf{i} + y\mathbf{j}$, then $curl\mathbf{v}(Q)$ changes its direction, thus always being normal to $\overrightarrow{OQ}$ but remaining in the $x - y$ plane. The magnitude of $curl\ \mathbf{v}(Q)$ is twice the magnitude of $\overrightarrow{OQ}$. If we place a small helix at point Q, with the plane of its blades coincident with the meridian plane $\theta = constant$ that contains Q, we can see it rotating, with the axis of rotation along $curl\ \mathbf{v}(Q)$. The orientation $\hat{\mathbf{k}} \times \hat{\mathbf{r}}$ can be intuitively felt from the fact that, keeping θ constant, if we move Q a little nearer the axis, the velocity increases and the helix blades nearer the axis will acquire a velocity in the axial direction greater than those that are farther away, thereby forcing the small helix to rotate with an angular velocity vector directed along $curl\ \mathbf{v}(Q)$.

Appendix G

Green's Identities

Gauss' divergence theorem has several useful, easy formulas, as corollaries which are collectively known as Green's identities.

Setting $\mathbf{v} = u\ grad\ f$, we get

$$\int\limits_{\Omega} div(u\ grad\ f)\ d\Omega = \int\limits_{\sigma} u\ grad\ f \cdot \mathbf{n}\ d\sigma = \int\limits_{\sigma} u\frac{df}{dn}d\sigma. \qquad (G.1)$$

By the same token:

$$\int\limits_{\Omega} div(f\ grad\ u)\ d\Omega = \int\limits_{\sigma} f\ grad\ u \cdot \mathbf{n}\ d\sigma = \int\limits_{\sigma} f\frac{du}{dn}d\sigma. \qquad (G.2)$$

However

$$div(u\ grad\ f) = u\nabla^2 f + grad\ u \cdot grad\ f \qquad (G.3)$$
$$div(f\ grad\ u) = f\nabla^2 u + grad\ f \cdot grad\ u, \qquad (G.4)$$

so, by subtracting the last two differential identities, we get

$$div(u\ grad\ f) - div(f\ grad\ u) = u\nabla^2 f - f\nabla^2 u, \qquad (G.5)$$

so that

$$\int\limits_{\Omega} (u\nabla^2 f - f\nabla^2 u)d\Omega = \int\limits_{\sigma} \left(u\frac{df}{dn} - f\frac{du}{dn} \right) d\sigma. \qquad (G.6)$$

The name Green is also associated with the so-called Green's theorem, which can be summarized in the following formula (known as the Green's formula for the 2-D case)

$$\int\int\limits_{\Omega} \left(\frac{\delta P}{\delta x} - \frac{\delta Q}{\delta y} \right)(x,y) = \int\limits_{P} (x,y)dx + Q(x,y)dy. \qquad (G.7)$$

As shown in Appendix D, this formula can be regarded as a particular case of the more general Stoke's formula, or as a particular case of Gauss' divergence theorem, when applied to a 3-D body of cylindrical shape with a body (of unit-height plane cross-section Ω) acting as a domain for the 2-D vector field $\mathbf{v} = Q\mathbf{i} - P\mathbf{j}$.

Appendix H

Fluid Volumetric Time-Tate of Change

For simplicity's sake, let us consider a compressible amount of fluid inside a tubular region $\Omega(t)$ with volume $V(t)$ at time t. $\Omega(t)$ has an input face σ_i with area A_i, and an output face σ_o with area $A_o = A_i$. The outwardly directed unit normal to σ_i and σ_o are $\mathbf{n}_i$ and $\mathbf{n}_o$, respectively. This can be visualized using images of oil or gas pipelines. The fluid is moving through such a pipe with a nonuniform velocity $\mathbf{v}(t)$ along its axis, but uniformly throughout each crosssection, so that particles are crossing σ_i and σ_o with velocities $\mathbf{v}_i$ and $\mathbf{v}_o$, respectively, at time t.

Consider the positions reached by the fluid particles that were between σ_i and σ_o at time t, after a time interval duration of Δt has elapsed (see Figure H.1.)

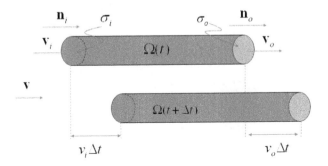

Figure H.1: Rectilinear tubular flow.

Consider:

$$V(t) = \int_{\Omega(t)} 1 \, dV. \tag{H.1}$$

Let $\mathbf{u}$ be a unit vector in the flow direction, so that $\mathbf{v}_i = v_i \mathbf{u}$, $\mathbf{v}_o = v_o \mathbf{u}$, $\mathbf{n}_o = \mathbf{u}$ and $\mathbf{n}_i = -\mathbf{u}$:

$$V(t + \delta t) \approx V(t) + A_o v_o \Delta t - A_i v_i \Delta t, \tag{H.2}$$

203

so that

$$\frac{V(t+\Delta t) - V(t)}{\Delta t} \approx A_o v_o - A_i v_i = A_o \mathbf{v}_o \cdot \mathbf{n}_o + A_i \mathbf{v}_i \cdot \mathbf{n}_i = \int_{\sigma(t)} \mathbf{v} \cdot \mathbf{n}\, d\sigma. \quad \text{(H.3)}$$

Taking the limit as Δt tends to zero

$$\frac{dV(t)}{dt}(t) = \int_{\sigma(t)} \mathbf{v} \cdot \mathbf{n}\, d\sigma = \int_{\Omega(t)} div\ \mathbf{v}\, dV, \quad \text{(H.4)}$$

where $\sigma(t)$ is the surface bounding the cylinder $\Omega(t)$. This result can be rewritten as

$$\frac{d}{dt} \int_{\Omega(t)} 1 dV = \int_{\Omega(t)} div\ \mathbf{v}\, dV. \quad \text{(H.5)}$$

Next, we imitate Faraday's notion of "flux tubes" and consider a fluid flow constituted by curved trajectories or "streamlines," i.e., curved lines which are everywhere tangent to the velocity field $\mathbf{v}$. We also assume that $\mathbf{v}$ is uniform throughout each cross-section of the "tube" formed by a collection of neighboring streamlines (see Figure H.2.)

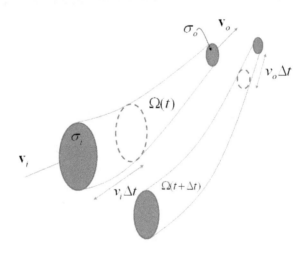

Figure H.2: Faraday flux tubes.

With the same obvious notations as in the rectilinear flow, it is immediately obvious that, for flux tubes with small cross-sections σ_i and σ_o, the following is a good approximation for a small Δt:

$$\frac{V(t+\Delta t) - V(t)}{\Delta t} \approx A_o v_o - A_i v_i = A_o \mathbf{v}_o \cdot \mathbf{n}_o + A_i \mathbf{v}_i \cdot \mathbf{n}_i = \int_{\sigma(t)} \mathbf{v} \cdot \mathbf{n}\, d\sigma, \quad \text{(H.6)}$$

where $\sigma(t)$ is the surface bounding the flux tube $\Sigma(t)$. Obviously, $\sigma(t) = \sigma_i \cup M(t) \cup \sigma_i$, where $M(t)$ denotes "the mantle" of $\Omega(t)$, which is a space portion where $\mathbf{v} \cdot \mathbf{n} = 0$, by definition of a "flux tube".

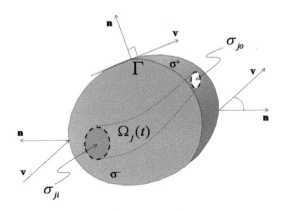

Figure H.3: Faraday flux tube through volume Ω.

For a general case, we reproduce Figure H.1, but apply to it an analysis-synthesis procedure. We decompose σ^-, the portion of $\sigma = \partial\Omega$ where $\mathbf{v}\cdot\mathbf{n} = 0$, in a large number of sufficiently small surface elements σ_{ji}, $j = 1, ..., N$, so that regarding them as planar is a good approximation. A flux tube is then built up with the flow streamlines incoming through σ_{ji}, into the 3-D region Ω bounded by $\sigma = \sigma^+ \cup \sigma^- \cup \Gamma$. These tube streamlines have an output flux through $\sigma_{jo} \subset \sigma^+$, $j = 1, ..., N$. The "wormhole" looking flux tube Ω_j starting from σ_{ji} and ending at σ_{jo}, has a small volume $V_j(t)$ (See Figure H.3), and, obviously:

$$V(t) = \int_{\Omega(t)} 1\,dV = \sum_{j=1}^{N} V_j(t), \tag{H.7}$$

so

$$\frac{V(t + \Delta t) - V(t)}{\Delta t} \approx \sum_{j=1}^{N} \frac{V_j(t + \Delta t) - V_j(t)}{\Delta t}. \tag{H.8}$$

Applying our previous result to each flux tube Ω_j, we have

$$\frac{V_j(t + \Delta t) - V_j(t)}{\Delta t} \approx A_{jo}\mathbf{v}_{jo} \cdot \mathbf{n}_{jo} + A_{ji}\mathbf{v}_{ji} \cdot \mathbf{n}_{ji}, \tag{H.9}$$

where the labels "o" and "i" stand for "output" and "input," respectively.

Therefore:

$$\frac{V(t + \Delta t) - V(t)}{\Delta t} \approx \sum_{j=1}^{N} A_{jo} \mathbf{v}_{jo} \cdot \mathbf{n}_{jo} + \sum_{j=1}^{N} A_{ji} \mathbf{v}_{ji} \cdot \mathbf{n}_{ji} \qquad \text{(H.10)}$$

$$= \int_{\sigma+} \mathbf{v} \cdot \mathbf{n} + \int_{\sigma-} \mathbf{v} \cdot \mathbf{n} = \int_{\sigma} \mathbf{v} \cdot \mathbf{n}. \qquad \text{(H.11)}$$

Taking the limit as Δt tends to zero, and using Gauss' divergence formula, we get

$$\frac{dV(t)}{dt} = \int_{\Omega(t)} div\ \mathbf{v} dV, \qquad \text{(H.12)}$$

or

$$\frac{dV(t)}{dt} - \int_{\Omega(t)} div\ \mathbf{v} dV = 0. \qquad \text{(H.13)}$$

This formula should be compared with the continuity equation given in §2.3.1; when compared, the sign difference observed here is easily understandable, since for a given mass, a volume increment will be followed by a density increment.

Appendix I

General Formulation of the Flux Concept

Let us recall that the natural notion of **mass flow** or "flux" $\dot{M}(\sigma)$ of fluid particles across some smooth surface σ, from west-to-east. The flow forced by a velocity field $\mathbf{x}(x,t)$, imparted to particles located at point x in time t, can be mathematically formulated by an analysis-synthesis two-stage process, resulting in the following definitory flux formula for fluids with mass density ρ and surfaces with a continuous field of unit normals, which have a west-to east-orientation, thus yielding

$$F_{we}(\sigma) = \text{ flux } = \dot{M}(\sigma) = \int_{\sigma} \rho \mathbf{v} \cdot \mathbf{n}_{we} d\sigma. \tag{I.1}$$

The vector field $\rho \mathbf{v}$ acts as a "vector density" for this "flux" F.

There are many vector fields describing physical vector magnitudes, for which corresponding "fluxes" can be analogously defined.

Starting with some planar surface σ, with a unit normal $\mathbf{n}_{we}$ and oriented from west to east, consider a uniform vector field $\mathbf{j} = J\mathbf{n}_{we}$. The flux $F_{we}(\sigma)$ of $\mathbf{j}$ across σ from west to east is naturally defined as

$$F_{we}(\sigma) = J(\mathbf{u} \cdot \mathbf{n}_{we}) \int_{\sigma} d\sigma = \int_{\sigma} (J\mathbf{u}) \cdot \mathbf{n}_{we} d\sigma, \tag{I.2}$$

and the corresponding **vector flux density** will be $\mathbf{j} = J\mathbf{u}$.

When the field $\mathbf{j}$ is nonuniform, both in size and direction, and σ is a smooth orientable surface, then the consideration of σ as the limit of polyhedral surfaces with increasingly small plane surface elements, lead us to define the flux as:

$$F_{we}(\sigma) = \int_{\sigma} \mathbf{j} \cdot \mathbf{n}_{we} \, d\sigma. \tag{I.3}$$

Appendix J

Fourth-Order Castillo–Grone Divergence Operators

For each given even order of accuracy k, there is a stencil $s = [s_1, s_2, ..., s_k]$ that fills the "Toeplitz-type" structure on the interior rows of the desired matrix $\mathbf{D}$, with bandwidth $b = k$. To modify the simplest matrix $\mathbf{S_D}$ to obtain a higher-order approximation on the boundary, we only need to modify the upper left and lower right corners of $\mathbf{S_D}$.

We define A as a t-by-l matrix, $A' = P_t A P_l$, with $t = k$ and $l = (3/2)k$. Therefore, the general form of our desired $h\mathbf{D} = h\mathbf{D}(A)$ will look like

$$
h \begin{bmatrix}
 & 0 \cdots & & & \cdots 0 \\
 & A \; 0 \cdots & & & \cdots 0 \\
 & 0 \cdots & & & \cdots 0 \\
 0 \cdots 0 & s_1 \; s_2 \; \cdots \; s_k & 0 \; 0 \; 0 \cdots 0 \\
 0 \cdots 0 \; 0 & \ddots \; \ddots \; \cdots \; \ddots & 0 \; 0 \cdots 0 \\
 0 \cdots 0 \; 0 & 0 \; s_1 \; s_2 \; \cdots \; s_k \; 0 \cdots 0 \\
 0 \cdots & & & \cdots 0 \\
 0 \cdots & & & \cdots 0 \quad A' \\
 0 \cdots & & & \cdots 0
\end{bmatrix} . \tag{J.1}
$$

The nonzero portions of A and A' in $h\mathbf{D}(A)$ do not overlap, since we imposed the condition

$$
N \geq 2l - 1 = 3k - 1 = 3b - 1. \tag{J.2}
$$

Since A is a k-by-$(3/2)k$ matrix, then for $k = 4$ we have $A \in \mathbb{R}^{(4 \times 6)}$.

Let $a_i = row_i(A)$, so that $a_1 = [a_{11} \; a_{12} \; a_{13} \; a_{14} \; a_{15} \; a_{16}]$, and so on.

The conditions on A will be described by a matrix equation. $\mathbf{D}(A)$ must satisfy the row sum, column sum, and order constraints, and when we introduce a Vandermonde matrix V_i, the desired conditions on a_i for $k = 4$ can be expressed by the four matrix equations:

$$
V_1 a_1^T = V_2 a_2^T = V_3 a_3^T = V_4 a_4^T = [0 \; 1 \; 0 \; 0 \; 0]^T, \tag{J.3}
$$

which establish five scalar constraints on each a_i, $i = 1, ..., 4$.

The first constraint is that the row sum of a_i is zero. The next four describe the conditions on a_i to obtain a fourth-order approximation.

A general **Vandermonde matrix** $V(m; g_1, ..., g_n)$ is defined as an $(m+1)$-by-n matrix:

$$\begin{bmatrix} 1 & \cdots & 1 \\ g_1 & \cdots & g_n \\ g_1^2 & \cdots & g_n^2 \\ \vdots & \cdots & \vdots \\ g_1^m & \cdots & g_n^m \end{bmatrix}. \tag{J.4}$$

The row $[g_1 \ ... \ g_n]$ is called the **generator** of V.

In the present situation of desired fourth-order accuracy, $m = 4$ and $n = (3/2)k = 6$.

The first Vandermonde matrix, $V_1 = V(4; -1/2, 1/2, 3/2, 5/2, 7/2, 9/2)$, allows us to construct V_2, V_3, and V_4 by simply subtracting the quantity 1 each time to the six values $g_1, ..., g_6$ of $V_1 = V(4; g_1, ..., g_6)$:

$$V_2 = V(4; -3/2, -1/2, 1/2, 3/2, 5/2, 7/2), \tag{J.5}$$
$$V_3 = V(4; -5/2, -3/2, -1/2, 1/2, 3/2, 5/2), \tag{J.6}$$
$$V_4 = V(4; -7/2, -5/2, -3/2, -1/2, 1/2, 3/2). \tag{J.7}$$

We see that the generator for V_1 is $(-1/2, 1/2, 3/2, 5/2, 7/2, 9/2)$, and can now seek for the solution of $V_1 a_1^T = [0 \ 1 \ 0 \ 0 \ 0]^T$; that is

$$\begin{bmatrix} 1 & 1 & 1 & 1 & 1 & 1 \\ -1/2 & 1/2 & 3/2 & 5/2 & 7/2 & 9/2 \\ (-1/2)^2 & (1/2)^2 & (3/2)^2 & (5/2)^2 & (7/2)^2 & (9/2)^2 \\ (-1/2)^3 & (1/2)^3 & (3/2)^3 & (5/2)^3 & (7/2)^3 & (9/2)^3 \\ (-1/2)^4 & (1/2)^4 & (3/2)^4 & (5/2)^4 & (7/2)^4 & (9/2)^4 \end{bmatrix} \begin{bmatrix} a_{11} \\ a_{12} \\ a_{13} \\ a_{14} \\ a_{15} \\ a_{16} \end{bmatrix} = \begin{bmatrix} 0 \\ 1 \\ 0 \\ 0 \\ 0 \end{bmatrix}. \tag{J.8}$$

We also look for the solutions a_2, a_3, and a_4 to equations $V_2 a_2^T = V_3 a_3^T = V_3 a_3^T = [0 \ 1 \ 0 \ 0 \ 0]^T$. The solutions a_1, a_2, a_3, and a_4 to these equations are not unique. In general, the null space of a k-th order Vandermonde matrix V_i ($i = 1, ..., 4$) has a dimension $((1/2)k - 1)$, so that, for $k = 4$, the four Vandermonde matrices $v_1, ..., v_4$ all have a one-dimensional null space in $\mathbb{R}^6$, which turns out to be the same for all of them, and is generated by the six-dimensional vector $[-1 \ 5 \ - \ 10 \ 10 \ - \ 5 \ 1]^T$, as can be easily verified.

Having found that the null space for $V(4; g_1, ..., g_6)$ is one dimensional, having a different generator for each V_i, $i = 1, ...4$, and using easily found particular solutions of the nonhomogeneous linear algebraic systems $V_i(a_i)^T = [0 \ 1 \ 0 \ 0 \ 0]^T$, $i = 1, ...4$, we can write the general solutions for a_1, a_2, a_3, and

a_4, as follows, with α_1, α_2, α_3, and α_4 real parameters:

$$a_1 = row_1(A) = \begin{bmatrix} -\dfrac{11}{12} & \dfrac{17}{24} & \dfrac{3}{8} & -\dfrac{5}{24} & \dfrac{1}{24} & 0 \end{bmatrix} + \alpha_1[-1\ 5\ -10\ 10\ -5\ 1], \text{(J.9)}$$

$$a_2 = row_2(A) = \begin{bmatrix} \dfrac{1}{24} & -\dfrac{9}{8} & \dfrac{9}{8} & -\dfrac{1}{24} & 0 & 0 \end{bmatrix} + \alpha_2[-1\ 5\ -10\ 10\ -5\ 1], \text{(J.10)}$$

$$a_3 = row_3(A) = \begin{bmatrix} 0 & \dfrac{1}{24} & -\dfrac{9}{8} & \dfrac{9}{8} & -\dfrac{1}{24} & 0 \end{bmatrix} + \alpha_3[-1\ 5\ -10\ 10\ -5\ 1], \text{(J.11)}$$

$$a_4 = row_4(A) = \begin{bmatrix} 0 & 0 & \dfrac{1}{24} & -\dfrac{9}{8} & \dfrac{9}{8} & -\dfrac{1}{24} \end{bmatrix} + \alpha_4[-1\ 5\ -10\ 10\ -5\ 1]. \text{(J.12)}$$

Designating the common null space vector as $\nu \equiv [-1\ 5\ -10\ 10\ -5\ 1]^T$ and $\alpha = (\alpha_1,\ \alpha_2,\ \alpha_3,\ \alpha_4)$, our desired matrix $A(\alpha)$ can be written in a compact form

$$A(\alpha) = \Pi + \alpha\nu^T = \begin{bmatrix} -\dfrac{11}{12} & \dfrac{17}{24} & \dfrac{3}{8} & -\dfrac{5}{24} & \dfrac{1}{24} & 0 \\ \dfrac{1}{24} & -\dfrac{9}{8} & \dfrac{9}{8} & -\dfrac{1}{24} & 0 & 0 \\ 0 & \dfrac{1}{24} & -\dfrac{9}{8} & \dfrac{9}{8} & -\dfrac{1}{24} & 0 \\ 0 & 0 & \dfrac{1}{24} & -\dfrac{9}{8} & \dfrac{9}{8} & -\dfrac{1}{24} \end{bmatrix} +$$

$$\begin{bmatrix} \alpha_1 \\ \alpha_2 \\ \alpha_3 \\ \alpha_4 \end{bmatrix} [-1\ 5\ -10\ 10\ -5\ 1]. \tag{J.13}$$

Up to this point, $\mathbf{D}(A(\alpha))$ is a four parameter family of fourth-order accuracy everywhere, mapping from $\mathbb{R}^{(n+1)}$ to $\mathbb{R}^n$, as we will prove below.

Consider a grid position x, and four surrounding grid points of the form $(x - (3/2)h)$, $(x - (1/2)h)$, $(x + (1/2)h)$, and $(x + (3/2)h)$. If we want to approximate $f'(x)$ in terms of $f(x - (3/2)h)$, $f(x - (1/2)h)$, $f(x + (1/2)h)$, and $f(x + (3/2)h)$ with local truncation error of $O(h^4)$, then it is necessary to find four coefficients $\sigma_1, ..., \sigma_4$, such that:

$$\sigma_1 f(x - \frac{3}{2}h) + \sigma_2 f(x - \frac{1}{2}h) + \sigma_3 f(x + \frac{1}{2}h) + \sigma_4 f(x + \frac{3}{2}h) = f'(x) + O(h^4). \tag{J.14}$$

Therefore, if we define the 1-by-4 matrix $\sigma = [\sigma_1\ \sigma_2\ \sigma_3\ \sigma_4]$, then

$$\sigma \begin{bmatrix} f(x - \frac{3}{2}h) \\ f(x - \frac{1}{2}h) \\ f(x + \frac{1}{2}h) \\ f(x + \frac{3}{2}h) \end{bmatrix} = f'(x) + O(h^4). \tag{J.15}$$

On the other hand, using Taylor expansions, it is easy to see that

$$
\begin{bmatrix} f(x-\frac{3}{2}h) \\ f(x-\frac{1}{2}h) \\ f(x+\frac{1}{2}h) \\ f(x+\frac{3}{2}h) \end{bmatrix} - \begin{bmatrix} O(h^5) \\ O(h^5) \\ O(h^5) \\ O(h^5) \end{bmatrix} =
$$

$$
\begin{bmatrix} f(x) - \frac{3}{2}hf'(x) + \left(\frac{3}{2}h\right)^2 \frac{f''(x)}{2!} - \left(\frac{3}{2}h\right)^3 \frac{f''3(x)}{3!} + \left(\frac{3}{2}h\right)^4 \frac{f^{IV}(x)}{4!} \\ f(x) - \frac{1}{2}hf'(x) + \left(\frac{1}{2}h\right)^2 \frac{f''(x)}{2!} - \left(\frac{1}{2}h\right)^3 \frac{f''3(x)}{3!} + \left(\frac{1}{2}h\right)^4 \frac{f^{IV}(x)}{4!} \\ f(x) + \frac{1}{2}hf'(x) + \left(\frac{1}{2}h\right)^2 \frac{f''(x)}{2!} + \left(\frac{1}{2}h\right)^3 \frac{f''3(x)}{3!} + \left(\frac{1}{2}h\right)^4 \frac{f^{IV}(x)}{4!} \\ f(x) + \frac{3}{2}hf'(x) + \left(\frac{3}{2}h\right)^2 \frac{f''(x)}{2!} + \left(\frac{3}{2}h\right)^3 \frac{f''3(x)}{3!} + \left(\frac{3}{2}h\right)^4 \frac{f^{IV}(x)}{4!} \end{bmatrix} . \quad \text{(J.16)}
$$

This can be rewritten using a Vandermonde matrix V, as follows:

$$
\begin{bmatrix} f(x-\frac{3}{2}h) \\ f(x-\frac{1}{2}h) \\ f(x+\frac{1}{2}h) \\ f(x+\frac{3}{2}h) \end{bmatrix} = \begin{bmatrix} 1 & -\frac{3}{2} & \left(-\frac{3}{2}\right)^2 & \left(-\frac{3}{2}\right)^3 & \left(-\frac{3}{2}h\right)^4 \\ 1 & -\frac{1}{2} & \left(-\frac{1}{2}\right)^2 & \left(-\frac{1}{2}\right)^3 & \left(-\frac{1}{2}\right)^4 \\ 1 & \frac{1}{2} & \left(\frac{1}{2}\right)^2 & \left(\frac{1}{2}\right)^3 & \left(\frac{1}{2}\right)^4 \\ 1 & \frac{3}{2} & \left(\frac{3}{2}\right)^2 & \left(\frac{3}{2}\right)^3 & \left(\frac{3}{2}\right)^4 \end{bmatrix} \begin{bmatrix} f(x) \\ hf'(x) \\ h^2 \frac{f''(x)}{2!} \\ h^3 \frac{f'''(x)}{3!} \\ h^4 \frac{f^{IV}(x)}{4!} \end{bmatrix} + \begin{bmatrix} O(h^5) \\ O(h^5) \\ O(h^5) \\ O(h^5) \end{bmatrix}
$$

$$
\triangleq V^T \begin{bmatrix} f(x) \\ hf'(x) \\ h^2 \frac{f''(x)}{2!} \\ h^3 \frac{f'''(x)}{3!} \\ h^4 \frac{f^{IV}(x)}{4!} \end{bmatrix} + \begin{bmatrix} O(h^5) \\ O(h^5) \\ O(h^5) \\ O(h^5) \end{bmatrix} . \quad \text{(J.17)}
$$

Next, multiply this 4-by-1 matrix equation on the left by the 1-by-4 matrix σ to get

$$
f'(x) + \mathscr{O}(h^4) = \sigma V^T \begin{bmatrix} f(x) \\ hf'(x) \\ h^2 \frac{f''(x)}{2!} \\ h^3 \frac{f'''(x)}{3!} \\ h^4 \frac{f^{IV}(x)}{4!} \end{bmatrix} + O(h^5)(\sigma_1 + \sigma_2 + \sigma_3 + \sigma_4). \quad \text{(J.18)}
$$

Since $f'(x) = (1/h)[hf'(x)]$, then, while approximating $f'(x)$ with the linear combination $\sigma_1 f(x - \frac{3}{2}h) + \sigma_2 f(x - \frac{1}{2}h) + \sigma_3 f(x + \frac{1}{2}h) + \sigma_4 f(x + \frac{3}{2}h) = f'(x) + O(h^4)$, we require $\sigma = \sigma(h)$ to be such that

$$
\sigma V^T \triangleq [0 \; 1/h \; 0; 0; 0], \quad \text{(J.19)}
$$

ensuring that

$$\sigma V^T \begin{bmatrix} f(x) \\ hf'(x) \\ h^2\frac{f''(x)}{2!} \\ h^3\frac{f'''(x)}{3!} \\ h^4\frac{f^{IV}(x)}{4!} \end{bmatrix} = [0\ 1/h\ 0; 0; 0] \begin{bmatrix} f(x) \\ hf'(x) \\ h^2\frac{f''(x)}{2!} \\ h^3\frac{f'''(x)}{3!} \\ h^4\frac{f^{IV}(x)}{4!} \end{bmatrix} = (1/h)[hf'(x)] = f'(x).$$

(J.20)

We see that to keep terms in both sides of equation (J.18) as magnitudes of the same order with respect to h, σ must be proportional to $(1/h)$, so that we define an auxiliary coefficient 1-by-4 matrix $s = [s_1\ s_2\ s_3\ s_4]$, now independent of h, and such that $\sigma_j = s_j/h$, $j = 1, ..., 4$, implying $O(h^5)(\sigma_1+\sigma_2+\sigma_3+\sigma_4) = O(h^4)(s_1+s_2+s_3+s_4)$, and (J.18) is satisfied now, if and only if (J.19) holds; namely:

$$\left(\frac{s}{h}\right)V^T = [0\ (1/h)\ 0\ 0\ 0],$$ (J.21)

or

$$Vs^T = [0\ 1\ 0\ 0\ 0]^T.$$ (J.22)

This resulting equation is easily solved for

$$[s_1\ s_2\ s_3\ s_4] = \left[\frac{1}{24}\ -\frac{9}{8}\ \frac{9}{8}\ -\frac{1}{24}\right],$$ (J.23)

which is exactly the 4th-order centered finite difference stencil.

Analogously, using Taylor's series and Vandermonde matrices will yield 4th-order accurate Castillo–Grone divergence near and at the boundary points.

Now that we have constructed the basic stencil

$$s = [s_1\ s_2\ s_3\ s_4] = \left[\frac{1}{24}\ -\frac{9}{8}\ \frac{9}{8}\ -\frac{1}{24}\right],$$ (J.24)

we can take a closer look at the upper left corner of $h\mathbf{D}(A)$:

$$h\mathbf{D}(A) = \begin{bmatrix} a_{11}\ a_{12}\ a_{13}\ a_{14}\ a_{15}\ a_{16} & 0 & \cdots \\ a_{21}\ a_{22}\ a_{23}\ a_{24}\ a_{25}\ a_{26} & 0 & \cdots \\ a_{31}\ a_{32}\ a_{33}\ a_{34}\ a_{35}\ a_{36} & 0 & \cdots \\ a_{41}\ a_{42}\ a_{43}\ a_{44}\ a_{45}\ a_{46} & 0 & \cdots \\ 0\ 0\ 0\ \frac{1}{24}\ -\frac{9}{8}\ \frac{9}{8}\ -\frac{1}{24}\ 0\ \cdots \\ \vdots\ \vdots\ \vdots\ 0\ \frac{1}{24}\ -\frac{9}{8}\ \frac{9}{8}\ -\frac{1}{24}\ 0\ \cdots \\ \vdots\ 0\ \frac{1}{24}\ -\frac{9}{8}\ \frac{9}{8}\ -\frac{1}{24}\ 0\ \cdots \\ \vdots\ 0\ \frac{1}{24}\ -\frac{9}{8}\ \frac{9}{8}\ -\frac{1}{24}\ 0\ \cdots \\ \vdots\ 0\ \ddots \\ \vdots \end{bmatrix}$$

We see that the column sum for the fourth column is $(a_{14} + a_{24} + a_{34} + a_{44} + (1/24))$, and for the fifth column is $(a_{15} + a_{25} + a_{35} + a_{45} - (26/24))$, and so on.

Having exhibited a typical accuracy analysis with the help of Vandermonde matrices, we next address our attention to choose, among this large 4-parameter of 4th-order accurate divergence operators $\mathbf{D}(A(\alpha))$, **only those members that are mimetic.**

Obviously, the derivative of a constant function is 0 for all values of α, since the row sum for $a_1(\alpha_1)$, $a_2(\alpha_2)$, $a_3(\alpha_3)$ and $a_4(\alpha_4)$ is zero.

In order to satisfy the fundamental theorem of calculus condition, we require that the column sums of $h\mathbf{D}(A)$ to equal $\{-1, 0, ..., 1\}$.

Now,

$$1_4^T A(\alpha) = [1\ 1\ 1\ 1]A(\alpha) = \left[\sum_{i=1}^{4} a_{i1}, \sum_{i=1}^{4} a_{i2}, \sum_{i=1}^{4} a_{i3}, ..., \sum_{i=1}^{4} a_{i6}\right]. \quad (J.25)$$

Looking at the upper left corner of the matrix $h\mathbf{D}(A)$, and at the 4th, 5th, and 6th column sums, we see that α must be chosen so that

$$1_4^T A(\alpha) = [-1\ 0\ 0\ -1/24\ 13/12\ -1/24] \triangleq d^T. \quad (J.26)$$

Now, it is easy to see that it is impossible to find α, such that $h\mathbf{D}(A(\alpha))$ satisfies the discrete form of the fundamental theorem of calculus for the standard inner product, which is where the N-by-N weight matrix

$$Q = DIAG\,(q_1, q_2, ..., q_k, 1, ..., 1, q_k, q_{k-1}, ..., q_1) \quad (J.27)$$

comes in.

J.1 Computing the Weight Matrix Q

Therefore, instead of solving $1_4^T A(\alpha) = d^T$, which is impossible, we now attempt to solve $1_N^T Q h\mathbf{D}(A(\alpha)) = [-1, 0, ..., 0, 1]$, which is the same as solving for q the reexpressed mimetic condition employing q:

$$1_4^T DIAG(q)A(\alpha) = d^T = [-1\ 0\ 0\ -1/24\ 13/12\ -1/24]. \quad (J.28)$$

Since $1_4^T DIAG(q) = q^T$, it only remains to solve $[A(\alpha)]^T q = d$.

Rearranging this equation into a form where all of the unknowns are in a single vector, we find

$$[A(\alpha)]^T q \equiv (\Pi^T + \nu\alpha^T)q = \Pi^T q + \nu(\alpha^T q), \quad (J.29)$$

but $(\alpha^T q)$ is a scalar λ being the product of 1-by-4 and 4-by-1 matrices, so that $\nu(\alpha^T q) = (\alpha^T q)\nu = \lambda\nu$, and the system to be solved becomes

$$\Pi^T q + \lambda\nu = d. \qquad (J.30)$$

Let us define $\tilde{q}^T = [q_1 \; q_2 \; q_3 \; q_4 \; \lambda]$ and $\Pi^T = \{p_{ij}\}$, so that

$$(\Pi^T q)_i = \sum_{j=1}^{4} p_{ij}q_j, \; i = 1, .., 6. \qquad (J.31)$$

Let $\tilde{\Pi}$ be a 6-by-5 matrix defined, as follows:

$$(\tilde{\Pi})_{ij} = p_{ij} \text{ for } 1 \le j \le 4, \qquad (J.32)$$

and $(\tilde{\Pi})_{i5} = \nu_i$, $i = 1, ..., 6$, so that

$$(\tilde{\Pi}\tilde{q})_i = \sum_{j=1}^{4} p_{ij}q_j + \nu_i\lambda = (\Pi^T q + \lambda\nu)_i = d_i, \; i = 1, ..., 6. \qquad (J.33)$$

Our unknowns are now q_1, q_2, q_3, q_4, and λ, and the system for these five unknowns is $\tilde{\Pi}\tilde{q} = d$, or, written in full:

$$
\begin{bmatrix}
-\frac{11}{12} & \frac{1}{24} & 0 & 0 & -1 \\
\frac{17}{24} & -\frac{9}{8} & \frac{1}{24} & 0 & 5 \\
\frac{3}{8} & \frac{9}{8} & -\frac{9}{8} & \frac{1}{24} & -10 \\
-\frac{5}{24} & -\frac{1}{24} & \frac{9}{8} & -\frac{9}{8} & 10 \\
\frac{1}{24} & 0 & -\frac{1}{24} & \frac{9}{8} & -5 \\
0 & 0 & 0 & -\frac{1}{24} & 1
\end{bmatrix}
\begin{bmatrix} q_1 \\ q_2 \\ q_3 \\ q_4 \\ \lambda \end{bmatrix}
=
\begin{bmatrix} -1 \\ 0 \\ 0 \\ -1/24 \\ 13/12 \\ -1/24 \end{bmatrix}.
$$

Fortunately, this system has a unique solution, because $\tilde{\Pi}$ has full rank:

$$\tilde{q}^T = [q_1 \; q_2 \; q_3 \; q_4 \; \lambda] = \left[\frac{649}{576} \; \frac{143}{192} \; \frac{75}{64} \; \frac{551}{576} \; -\frac{25}{13824}\right]. \qquad (J.34)$$

Therefore, we have obtained explicitly q, which determines the N-by-N weight matrix Q:

$$Q = DIAG\left(\frac{649}{576}, \frac{143}{192}, \frac{75}{64}, \frac{551}{576}, 1, ..., 1, \frac{551}{576}, \frac{75}{64}, \frac{143}{192}, \frac{649}{576}\right). \qquad (J.35)$$

We also have $\lambda = -25/13824$; therefore, since $\alpha^T q = \lambda$, this relationship can be written in full as follows:

$$\frac{649}{576}\alpha_1 + \frac{143}{192}\alpha_2 + \frac{75}{64}\alpha_3 + \frac{551}{576}\alpha_4 = -\frac{25}{13824}. \qquad (J.36)$$

This equation defines a 3-D hypersurface in $\mathbb{R}^4$, on which α must reside, so that $\mathbf{D}(A(\alpha))$ is a 3-parameter family, as stated before. In particular, taking $\alpha_2 = \alpha_3 = \alpha_4 = 0$, we get $\alpha^T q = (649/576)\alpha_1 = -25/13824$, which yields $\alpha_1 = -25/15576$, thus reducing the 4th-accurate divergence to the form

$$h\mathbf{D}(A) = h\check{\mathbf{D}}(A) = \begin{bmatrix} -\frac{4751}{5192} & \frac{909}{1298} & \frac{6091}{15576} & -\frac{1165}{5192} & \frac{129}{2596} & -\frac{25}{15576} & 0 & \cdots \\ \frac{1}{24} & -\frac{9}{8} & \frac{9}{8} & -\frac{1}{24} & 0 & 0 & 0 & \cdots \\ 0 & \frac{1}{24} & -\frac{9}{8} & \frac{9}{8} & -\frac{1}{24} & 0 & 0 & \cdots \\ 0 & 0 & \frac{1}{24} & -\frac{9}{8} & \frac{9}{8} & -\frac{1}{24} & 0 & \cdots \\ 0 & 0 & 0 & \frac{1}{24} & -\frac{9}{8} & \frac{9}{8} & -\frac{1}{24} & 0 & \cdots \\ 0 & 0 & 0 & 0 & \frac{1}{24} & -\frac{9}{8} & \frac{9}{8} & -\frac{1}{24} & 0 \end{bmatrix}$$

This is the same 4th-order DIV exemplified in [142].

The abovementioned particular choice of parameters has the advantage of exhibiting changes in the Castillo–Grone divergence operator, only in the first row of $\check{\mathbf{D}}(A)$ (second row of $\hat{\mathbf{D}}(A)$).

The technique employed here for the divergence is totally analogous to the one that would be employed for the gradient, and works for any even higher-order k.

References

[1] National Energy Technology Laboratory. Enhancing the Success of Carbon Capture and Storage Technologies. Technical report, U.S. Department of Energy, 2011.

[2] C. Di Bartolo, R. Gambini, and J. Pullin. Consistent and mimetic discretizations in general relativity. *J. Math. Phys.*, 46, 2005.

[3] A. DiCarlo, F. Milicchio, A. Paoluzzi, and V. Shapiro. Discrete physics using metrized chains. In *SPM '09: 2009 SIAM/ACM Joint Conference on Geometric and Physical Modeling*, pages 135–145, New York, NY, 2009. ACM.

[4] W. Chang, F. Giraldo, and B. Perot. Analysis of an exact fractional step method. *J. Comput. Phys.*, 180(1):183–199, 2002.

[5] K. Lipnikov, M. Shashkov, and D. Svyatskiy. The mimetic finite difference discretization of diffusion problem on unstructured polyhedral meshes. *J. Comput. Phys.*, 211(2):473–491, 2006.

[6] R. Liska, V. Ganzha, and C. Zenger. Mimetic finite difference methods for elliptic equations on unstructured grids. *Selcuk J. Appl. Math.*, 3(1):21–48, 2002.

[7] P.N. Vabishchevich. Finite-difference approximation of mathematical physics problems on irregular grids. *CMAM*, 5(3):294–330, 2005.

[8] M. Berndt, K. Lipnikov, P. Vachal, and M. Shashkov. A node reconnection algorithm for mimetic finite difference discretizations of elliptic equations on triangular meshes. *Commun. Math. Sci.*, 3(4):665–680, 2005.

[9] V. Ganzha, R. Liska, M. Shashkov, and C. Zenger. Support operator method for Laplace equation on unstructured triangular grid. *Selcuk J. Appl. Math.*, 3:21–48, 2002.

[10] R. Liska, M. Shashkov, and V. Ganzha. Analysis and optimization of inner products for mimetic finite difference methods on a triangular grid. *Math. Comput. Simulat.*, 67(1/2):55–66, 2004.

[11] F. Brezzi, K. Lipnikov, and M. Shashkov. Convergence of the mimetic finite difference method for diffusion problems on polyhedral meshes. *SIAM J. Numer. Anal.*, 43(5):1872–1896, 2005.

[12] A. Cangiani and G. Manzini. Flux reconstruction and solution post-processing in mimetic finite difference methods. *Comput. Method Appl. M.*, 197(9-12):933–945, 2008.

[13] V. Gyrya and K. Lipnikov. High-order mimetic finite difference method for diffusion problems on polygonal meshes. *J. Comput. Phys.*, 227(20):8841–8854, 2008.

[14] M. Vinokur. An analysis of finite-difference and finite-volume formulations of conservation laws. *J. Comput. Phys.*, 81(1):1–52, 1989.

[15] C. Mattiussi. The geometry of time-stepping. In Teixeira [76], pages 123–149.

[16] M. Bouman, A. Palha, J. Kreeft, and M. Gerritsma. A conservative spectral element method for curvilinear domains. In Hesthavena and Rønquist [186], pages 111–119.

[17] M. Gerritsma. Edge functions for spectral element methods. In Hesthavena and Rønquist [186], pages 199–207.

[18] M.I. Gerritsma, M. Bouman, and A. Palha. Least-squares spectral element method on a staggered grid. In I. Lirkov, S. Margenov, and J. Wasniewski, editors, *Lecture Notes in Computational Sience*, volume 5910, pages 659–666. Springer Verlag, 2010.

[19] J. Aarnes, S. Krogstad, and K-A Lie. Multiscale mixed/mimetic methods on corner-point grids. *Comput. Geosci.*, 12(3):297–315, 2008.

[20] L. Beirao da Veiga and G. Manzini. A higher-order formulation of the mimetic finite difference method. *SIAM J. Sci. Comput.*, 31(1):732–760, 2008.

[21] F. Hernandez, J.E. Castillo, and G.A. Larrazabal. Large sparse linear systems arising from mimetic discretization. *Comput. Math. Appl.*, 53:1–11, 2007.

[22] J.M. Hyman, J. Morel, M. Shashkov, and S. Steinberg. Mimetic finite difference methods for diffusion equations. *Comput. Geosci.*, 6(3):333–352, 2002. LA-UR-01-2434.

[23] B. Karasozen and V. Tsybulin. Conservative finite difference schemes for cosymmetric systems. In *Proc. 4th Conf. on Computer Algebra in Scientific Computing*, pages 363–375, Berlin, 2001. CASC, Springer.

[24] B. Karasozen and V.G. Tsybulin. Mimetic discretization of two-dimensional Darcy convection. *Comput. Phys. Comm.*, 167(3):203–213, 2005.

[25] Y. Kuznetsov, K. Lipnikov, and M. Shashkov. The mimetic finite difference method on polygonal meshes for diffusion-type problems. *Comput. Geosci.*, 8(4):301–324, 2004.

[26] O. Rojas, S.M. Day, J. Castillo, and L. Dalguer. Modeling of rupture propagation using high-order mimetic finite-differences. *Geophys. J. Int.*, 172:631–650, 2008.

[27] V.G. Tsybulin and B. Karasözen. Destruction of the family of steady states in the planar problem of Darcy convection. *Phys. Lett. A*, 372(35):5639–5643, 2008.

[28] M.F. Wheeler and I. Yotov. A multipoint flux mixed finite element method. *SIAM J. Numer. Anal.*, 44(5):2082–2106, 2006.

[29] A. Abbà and L. Bonaventura. A mimetic finite difference method for large eddy simulation of incompressible flow. Technical Report 34/2010, Politecnico di Milano, Milano, Italy, August 2010.

[30] E. Barbosa and O. Daube. A finite difference method for 3D incompressible flows in cylindrical coordinates. *Comput. Fluids*, 34(8):950–971, 2005.

[31] L. Bonaventura and T. Ringler. Analysis of discrete shallow water models on geodesic Delaunay grids with C-type staggering. *Mon. Weather Rev.*, 133(133):2351–2373, 2005.

[32] E. Haber and J. Modersitzki. A multilevel method for image registration. *SIAM J. Sci. Comp.*, 27(5):1594–1607, 2006.

[33] G. Yuan and Z. Sheng. Analysis of accuracy of a finite volume scheme for diffusion equations on distorted meshes. *J. Comput. Phys.*, 224(2):1170–1189, 2007.

[34] J. Yuan, C. Schorr, and G. Steidl. Simultaneous higher-order optical flow estimation and decomposition. *SIAM J. Sci. Comput.*, 29(6):2283–2304, 2007.

[35] C. Bazan, M. Abouali, J. Castillo, and P. Blomgren. Mimetic finite difference methods in image processing. *Comp. and Applied Mathematics*, 30(3):701–720, 2011.

[36] J.M. Hyman and M. Shashkov. Mimetic discretizations for Maxwell's equations and equations of magnetic diffusion. In J.A. DeSanto, editor, *Proc. of the Fourth International Conference on Mathematical and Numerical Aspects of Wave Propagation, Golden, Colorado, June 1-5, 1998*, pages 561–563, Philadelphia, 1998. SIAM.

[37] K.S. Yee. Numerical solution of initial boundary value problems involving maxwell's equations in isotropic media. *IEEE Trans. Antennas Propag.*, AP-14:302–307, 1966.

[38] K.S. Kunz and R.J. Luebbers. *The Finite Difference Time Domain Method for Electromagnetics*. CRC Press, Boca Raton, 1993.

[39] A. Bossavit. Geometrical methods in computational electromagnetism. In *Proceedings of ICAP 2006*, pages 1–6. Int. Comput. Accel. Phys. Conf., Joint Accelerator Conferences website, 2006.

[40] M. Clemens, P. Thoma, T. Weiland, and U. van Rienen. Computational electromagnetic-field calculation with the finite-integration method. *Surv. Math. Indust.*, 8:213–232, 1999.

[41] M. Clemens and T. Weiland. Discrete electromagnetism with the finite integration technique. In Teixeira [76], pages 65–87.

[42] M.E. Rose. Compact finite volume methods for the diffusion equation. *J. Sci. Comput.*, 3:261–290, 1989.

[43] B. Perot. Conservation properties of unstructured staggered mesh schemes. *J. Comput. Phys.*, 159(1):58–89, 2000.

[44] B. Perot. and R. Nallapati. A moving unstructured staggered mesh method for the simulation of incompressible free-surface flows. *J. Comput. Phys.*, 184(1):192–214, 2003.

[45] J.B. Perot, D. Vidovic, and P. Wesseling. Mimetic reconstruction of vectors, compatible spatial discretizations. In Arnold et al. [120], pages 173–188.

[46] J.B. Perot and V. Subramanian. Discrete calculus methods for diffusion. *J. Comput. Phys.*, 224(1):59–81, 2007.

[47] J.B. Perot and V. Subramanian. A discrete calculus analysis of the Keller box scheme and a generalization of the method to arbitrary meshes. *J. Comput. Phys.*, 226(1):494–508, 2007.

[48] R.J. LeVeque. High-resolution conservative algorithms for advection in incompressible flow. *SIAM J. Numer. Anal.*, 33(2):627–665, 1996.

[49] J. E. Morel, R. M. Roberts, and M. Shashkov. A local support-operators diffusion discretization scheme. *J. Comput. Phys.*, 144:17–51, 1998.

[50] M. Shashkov and S. Steinberg. Support-operator finite-difference algorithms for general elliptic problems. *J. Comput. Phys.*, 118(1):131–151, 1995.

[51] M. Shashkov and S. Steinberg. Solving diffusion equations with rough coefficients in rough grids. *J. Comput. Phys.*, 129(2):383–405, 1996.

[52] J.M. Hyman and M. Shashkov. Mimetic discretizations for Maxwell's equations. *J. Comput. Phys.*, 151(2):881–909, 1999.

[53] E.J. Caramana, D.E. Burton, M. Shashkov, and P.P. Whalen. The construction of compatible hydrodynamics algorithms utilizing conservation of total energy. *J. Comput. Phys.*, 146:227–262, 1998.

[54] J.M. Hyman and M. Shashkov. Mimetic finite difference methods for Maxwell's equations and the equations of magnetic diffusion. In Teixeira [76], pages 89–121.

[55] J.M. Hyman, M. Shashkov, and S. Steinberg. The numerical solution of diffusion problems in strongly heterogeneous non-isotropic materials. *J. Comput. Phys.*, 132(1):130–148, 1997.

[56] J.M. Hyman, M. Shashkov, and S. Steinberg. The effect of inner products for discrete vector fields on the accuracy of mimetic finite difference methods. *Comput. Math. Appl.*, 42(12):1527–1547, 2001.

[57] J.M. Hyman, R.J. Knapp, and J.C. Scovel. High order finite volume approximations of differential operators on nonuniform grids. *Physica D*, 60:112–138, 1992.

[58] J.M. Hyman and M. Shashkov. Adjoint operators for the natural discretizations of the divergence, gradient and curl on logically rectangular grids. *Appl. Numer. Math.*, 25(4):413–442, 1997.

[59] J. M. Hyman and M. Shashkov. Natural discretizations for the divergence, gradient, and curl on logically rectangular grids. *Comput. Math. Appl.*, 33(4):81–104, 1997.

[60] J.M. Hyman and M. Shashkov. The approximation of boundary conditions for mimetic finite difference methods. *Comput. Math. Appl.*, 36:79–99, 1998.

[61] J.M. Hyman and M. Shashkov. The orthogonal decomposition theorems for mimetic finite difference methods. *SIAM J. Numer. Anal.*, 36(3):788–818, 1999.

[62] J.M. Hyman, R.J. Knapp, and J.C. Scovel. High order finite volume approximations of differential operators on nonuniform grids. *Physica D*, 60(1-4):112–138, 1992.

[63] J.M. Hyman and J.C. Scovel. Deriving mimetic difference approximations to differential operators using algebraic topology. Technical report, Los Alamos National Laboratory, Los Alamos, NM, 1988.

[64] J.M. Hyman and S. Steinberg. The convergence of mimetic discretization for rough grids. *Comput. Math. Appl.*, 47(10-11):1565–1610, 2004.

[65] M. Shaskov. *Conservative Finite-Difference Methods on General Grids*. CRC Press, Boca Raton, FL, first edition, 1996.

[66] A.A. Samarskii, P.N. Vabishchevich, and P.P. Matus. *Difference Schemes with Operator Factors*. Kluwer Academic Publishers, Boston, 2002.

[67] A.A. Samarskii. *The Theory of Difference Schemes*, volume 240 of *Monographs and Textbooks in Pure and Applied Mathematics*. Marcel Dekker Inc., New York, NY, 2001. Translated from the Russian.

[68] A.A. Samarskii. *Teoriya raznostnykh skhem*. Izdat. "Nauka," Moscow, Russia, 1977.

[69] A.A. Samarskii, V.F. Tishkin, A.P. Favorskii, and M. Shashkov. Use of the support operator method for constructing difference analogues of operations of tensor analysis. *Differentsial nye Uravneniya*, 18(7):1251–1256, 1287, 1982. English translation: Differ. Equ. 18 (1982), no. 7, 881–885.

[70] N. Robidoux. *Numerical Solution of the Steady Diffusion Equation with Discontinuous Coefficients*. PhD thesis, University of New Mexico, Albuquerque, NM, May 2002.

[71] J.P. Zingano. *Convergence of Mimetic Methods for Sturm-Liouville Problems on General Grids*. PhD thesis, University of New Mexico, Albuquerque, NM, 2003.

[72] J.M. Guevara-Jordan, S. Rojas, M. Freites-Villegas, and J.E. Castillo. Convergence of a mimetic finite difference method for static diffusion equation. *Adv. Diff. Equat.*, 2007:1–12, 2007.

[73] J.M. Guevara-Jordan, S. Rojas, M. Freites-Villegas, and J.E. Castillo. A new second order finite difference conservative scheme. *Divulg. Mat.*, 13(1):107–122, 2005.

[74] J.C. Nédélec. Mixed finite elements in $\mathbf{r}^3$. *Numer. Math.*, 35(3):315–341, 1980.

[75] J.C. Nédélec. A new family of mixed finite elements in R^3. *Numer. Math.*, 50(1):57–81, 1986.

[76] F.L. Teixeira, editor. *Geometric Methods in Computational Electromagnetics, PIER 32*, Cambridge, Mass., 2001. EMW Publishing.

[77] F.L. Teixeira. Geometrical aspects of the simplicial discretization of Maxwell's equations. In *Geometric Methods in Computational Electromagnetics, PIER 32* [76], pages 171–188.

[78] F.L. Teixeira and W.C. Chew. Differential forms, metrics, and the reflectionless absorption of electromagnetic waves. *J. Electromagnet. Wave*, 13(5):665–686, 1999.

[79] F.L. Teixeira and W.C. Chew. Lattice electromagnetic theory from a topological viewpoint. *J. Math. Phys.*, 40(1):169–187, 1999.

[80] A. Bossavit. Mixed finite elements and the complex of Whitney forms. In *The Mathematics of Finite Elements and Applications, VI (Uxbridge, 1987)*, pages 137–144. Academic Press, London, 1988.

[81] A. Bossavit. Whitney forms: A new class of finite elements for three-dimensional computations in electromagnetics. *IEEE Proc. A*, 135:493–500, 1988.

[82] A. Bossavit. Differential forms and the computation of fields and forces in electromagnetism. *Eur. J. Mech. B-Fluid*, 10(5):474–488, 1991.

[83] A. Bossavit. Mixed methods and the marriage between "mixed" finite elements and boundary elements. *Numer. Meth. Part. D.E.*, 7(4):347–362, 1991.

[84] A. Bossavit. Edge-element computation of the force field in deformable bodies. *IEEE T. Magn.*, 28(2):1263–1266, 1992.

[85] A. Bossavit. On local computation of the force field in deformable bodies. *Int. J. Appl. Electrom. Mat.*, 2(4):333–343, 1992.

[86] A. Bossavit. *Computational electromagnetism*. Academic Press, San Diego, CA, 1998.

[87] A. Bossavit. How weak is the "weak solution" in finite element methods? *IEEE T. Magn.*, 34:2429–2432, 1998.

[88] A. Bossavit. The discrete Hodge operator in electromagnetic wave propagation problems. In *Mathematical and numerical aspects of wave propagation (Santiago de Compostela, 2000)*, pages 753–759. SIAM, Philadelphia, PA, 2000.

[89] A. Bossavit. Generalized finite differences in computational electromagnetics. In Teixeira [76], pages 45–64.

[90] A. Bossavit. Extrusion, contraction: their discretization via Whitney forms. *COMPEL*, 22(3):470–480, 2003.

[91] C. Mattiussi. An analysis of finite volume, finite element, and finite difference methods using some concepts from algebraic topology. *J. Comput. Phys.*, 133(2):289–309, 1997.

[92] C. Mattiussi. The finite volume, finite element, and finite difference methods as numerical methods for physical field problems. *Adv. Imag. Elect. Phys.*, 113:1–146, 2000.

[93] C. Mattiussi. A reference discretization strategy for the numerical solution of physical field problems. *Adv. Imag. Elect. Phys.*, 121:143–279, 2002.

[94] K. Mattsson. Boundary procedures for summation-by-parts operators. *SIAM J. Sci. Comput.*, 18(1):122–153, 2003.

[95] K. Mattsson. *Summation-by-parts operators for high order finite difference methods*. PhD thesis, Uppsala University, Uppsala, Sweden, 2003.

[96] K. Mattsson and J. Nordström. Summation by parts operators for finite difference approximations of second derivatives. *J. Comput. Phys.*, 199(2):503–540, 2004.

[97] P. Olsson. Summation by parts, projections, and stability i. *Math. Comput.*, 64-211:1035–1065, 1995.

[98] B. Strand. Summation by parts for finite difference approximations for d/dx. *J. Comput. Phys.*, 110:47–67, 1994.

[99] M. Svärd, K. Mattsson, and J. Nordström. Steady-state computations using summation-by-parts operators. *J. Sci. Comput.*, 24(1):79–95, 2005.

[100] H.O. Kreiss and G. Scherer. Finite element and finite difference methods for hyperbolic partial differential equations. In C. De Boor, editor, *Mathematical Aspects of Finite Elements in Partial Differential Equations*, pages 195–212, 1974.

[101] S. Steinberg. Applications of high-order discretizations to boundary-value problems. *CMAM*, 4(2):228–261, 2004.

[102] S. Steinberg. The accuracy of numerical models for continuum problems. In H. Bulgak and C. Zenger, editors, *Error Control and Adaptivity in Scientific Computing*, pages 299–323. NATO Science Series, 1999. Series C: Mathematical and Physical Sciences, 536.

[103] M.H. Carpenter, D. Gottlieb, and S. Abarbanel. Time-stable boundary conditions for finite difference solving hyperbolic systems: Methodology and applications to high-order compact schemes. *J. Comput. Phys.*, 111:220–236, 1994.

[104] M. Berndt, K. Lipnikov, J.D. Moulton, and M. Shashkov. Convergence of mimetic finite difference discretizations of the diffusion equation. *East-West J. Numer. Math.*, 9(4):253–316, 2001.

[105] F. Brezzi, K. Lipnikov, M.J. Shashkov, and V. Simoncinic. A new discretization methodology for diffusion problems on generalized polyhedral meshes. *Comput. Meth. Appl. M.*, 196(37-40):3682–3692, 2007.

[106] M. Berndt, K. Lipnikov, M.J. Shashkov, M.F. Wheeler, and I. Yotov. Superconvergence of the velocity in mimetic finite difference methods on quadrilaterals. *SIAM J. Numer. Anal.*, 43(4):1728–1749, 2005.

[107] J. Arteaga-Arispe and J.M. Guevara-Jordan. A conservative finite difference scheme for static diffusion equation. *Divulg. Mat.*, 16(1):39–54, 2008.

[108] S. Delcourte, K. Domelevo, and P. Omnes. A discrete duality finite volume approach to Hodge decomposition and div-curl problems on almost

arbitrary two-dimensional meshes. *SIAM J. Numer. Anal.*, 45(3):1142–1174, 2007.

[109] M. Desbrun, A.N. Hirani, M. Leok, and J.E. Marsden. Discrete Exterior Calculus. *ArXiv e-prints*, August 2005.

[110] E. Mansfield and P. Hydon. Difference forms. *Found. Comput. Math.*, 8(4):427–467, August 2008.

[111] N. Robidoux. A new method of construction of adjoint gradients and divergences on logically rectangular smooth grids. In F. Benkaldoun and R. Vilsmeier, editors, *Finite Volumes for Complex Applications: Problems and Perspectives, First International Symposium, July 15–18, 1996, Rouen, France*, pages 261–272, Paris, 1996. Éditions Hermès.

[112] A. Yavari. On geometric discretization of elasticity. *J. Math. Phys.*, 49(022901):1–36, 2008.

[113] N.M. Bessonov and D.J. Song. Application of vector calculus to numerical simulation of continuum mechanics problems. *J. Comput. Phys.*, 167:22–38, 2001.

[114] P. Castillo, R. Rieben, and D. White. Femster: An object-oriented class library of high-order discrete differential forms. *ACM T. Math. Software*, 31(4):425–457, 2005.

[115] J.A. Chard and V. Shapiro. A multivector data structure for differential forms and equations. *Math. Comput. Simulat.*, 54(1-3):33–64, 2000.

[116] P.W. Gross and P.R. Kotiuga. Data structures for geometric and topological aspects of finite element algorithms. In Teixeira [76], pages 151–169.

[117] R.S. Palmer. Chain models and finite element analysis: an executable CHAINS formulation of plane stress. *Comput. Aided Geom. D.*, 12(7):733–770, 1995.

[118] R.S. Palmer and V. Shapiro. Chain models of physical behavior for engineering analysis and design. *Res. Eng. Des.*, 5:161–184, 1993.

[119] W.M. Seiler. Completion to involution and semidiscretisations. *Appl. Numer. Math.*, 42(1):437–451, 2002.

[120] D.N. Arnold, P.B. Bochev, R. Lehoucq, R. Nicolaides, and M. Shashkov, editors. *Compatible discretizations: Proceedings of IMA Hot Topics workshop on Compatible discretizations*, volume 142 of *IMA Volumes in Mathematics and its Applications*, New York, NY, 2006. Springer.

[121] D.N. Arnold, R.S. Falk, and R. Winther. Finite element exterior calculus, homological techniques, and applications. *Acta Numer.*, 15(-1):1–155, 2006.

[122] P.B. Bochev and J. Hyman. Principles of mimetic discretizations of differential operators. In Arnold et al. [120], pages 89–120.

[123] S.H. Christiansen. A construction of spaces of compatible differential forms on cellular complexes. *Math. Mod. Meth. Appl. S.*, 18(5):739–757, 2008.

[124] A.T. de Hoop and I.E. Lager. Domain-integrated field approach to static magnetic field computation—application to some two-dimensional configurations. *IEEE T. Magn.*, 36(4):654–658, 2000.

[125] R. Hiptmair. Discrete Hodge-operators: an algebraic perspective. In Teixeira [76], pages 247–269.

[126] R. Hiptmair. High order Whitney forms. *J. Electromagnet. Wave*, 15(3):291–436, 2001.

[127] Z. Xie and H. Li. Exterior difference system on hypercubic lattice. *Acta Appl. Math.*, 99(1):97–116, 2007.

[128] H. Ammari and J.C. Nédélec. Couplage éléments finis/équations intégrales pour la résolution des équations de Maxwell en milieu hétérogène. In *Équations aux dérivées partielles et applications*, pages 19–33. Gauthier-Villars, Éd. Sci. Méd. Elsevier, Paris, 1998.

[129] P.B. Bochev, J. Hu, C.M. Siefert, and R.S. Tuminaro. An algebraic multigrid approach based on a compatible gauge reformulation of Maxwell's equations. *SIAM J. Sci. Comput.*, 31(1):557–583, 2009.

[130] F.H. Branin, Jr. The algebraic-topological basis for network analogies and for vector calculus. In *Symposium on Generalized Networks (12–14 April 1966)*, pages 453–491. Polytechnic Institute of Brooklyn, New York, 1966.

[131] V. Gradinaru and R. Hiptmair. Whitney elements on pyramids. *Electron. T. Numer. Anal.*, 8:154–168, 1999.

[132] P.W. Gross and P.R. Kotiuga. Finite element-based algorithms to make cuts for magnetic scalar potentials: topological constraints and computational complexity. In Teixeira [76], pages 207–245.

[133] R. Hiptmair. Higher order Whitney forms. In Teixeira [76], pages 271–299.

[134] R. Hiptmair. Finite elements in computational electromagnetism. *Acta Numer.*, 11:237–339, 2002.

[135] L. Kettunen, K. Forsman, and A. Bossavit. Gauging in Whitney spaces. *IEEE T. Magn.*, 35(3):1466–1469, May 1999.

[136] T. Tarhasaari and L. Kettunen. Topological approach to computational electromagnetism. In Teixeira [76], pages 189–206.

[137] T. Tarhasaari, L. Kettunen, and A. Bossavit. Some realizations of a discrete Hodge operator: a reinterpretation of finite element techniques. *IEEE T. Magn.*, 35:1494–1497, 1999.

[138] E. Tonti. The reason for analogies between physical theories. *Appl. Math. Model.*, 1(1):37–50, 1976/77.

[139] E. Tonti. On the geometrical structure of electromagnetism. In G. Ferraese, editor, *Gravitation, Electromagnetism and Geometrical Structures*, pages 281–308. Pitagora, Bologna, Italy, 1996.

[140] E. Tonti. Finite formulation of the electromagnetic field. In Teixeira [76], pages 1–44.

[141] E.M. von Hornbostel and C. Sachs. Systematik der musikinstrumente. *Ein Versuch Zeitschrift für Ethnologie*, 4/5, 1914.

[142] J.E. Castillo and R.D. Grone. A matrix analysis approach to higher-order approximations for divergence and gradients satisfying a global conservation law. *Siam J. Matrix Anal. Appl.*, 25:128–142, 2003.

[143] J.E. Castillo and M. Yasuda. Linear systems arising from second-order mimetic divergence and gradient discretizations. *J. of Math. Model. Algorithm*, 4:67–82, 2005.

[144] J.B. Runyan. A novel higher order finite difference time domain method based on the Castillo-Grone mimetic curl operator with application concerning the time-dependent Maxwell equations. Master's thesis, San Diego State University, San Diego, CA, 2011.

[145] J.M. Hyman, M.J. Shaskov, J.E. Castillo, and S. Steinberg. The sensitivity and accuracy of fourth order finite-difference schemes on non-uniform grids in one dimension. *Comput. Math. Appl.*, 30:41–55, 1995.

[146] A.L. Andrew. Centrosymmetric matrices. *SIAM Rev.*, 40:697–699, 1998.

[147] O. Montilla, C. Cadenas, and J.E. Castillo. Matrix approach to mimetic discretizations for differential operators on non-uniform grids. *Math. Comput. Simulation*, 73:215–225, 1993.

[148] J. Castillo, J. Hyman, M. Shashkov, and S. Steinberg. Fourth and sixth-order conservative finite difference approximations of the divergence and gradient. *Appl. Numer. Math.*, 37:171–187, 2001.

[149] I.A. Mannarino, J.M. Guevara, and Y. Quintana. A numerical study of a mimetic scheme for unsteady heat equation. *FARAUTE Ciens. y Tec.*, 3(1):64–70, 2008.

[150] E. Dijkstra. Go to statement considered harmful. *Communications of the ACM*, 11(3), 1968.

[151] M. Reddy. *API Design in C++*. Morgan Kauffmann, Massachusetts, first edition, 2011.

[152] C.L. Lawson, R.J. Hanson, D. Kincaid, , and F.T. Krogh. Basic linear algebra subprograms for fortran usage. *ACM Trans. Math. Soft*, 5:308–323, 1979.

[153] E. Anderson, Z. Bai, C. Bischof, S. Blackford, J. Demmel, J. Dongarra, J. Du Croz, A. Greenbaum, S. Hammarling, A. McKenney, and D. Sorensen. *LAPACK Users' Guide*. Society for Industrial and Applied Mathematics, Philadelphia, PA, third edition, 1999.

[154] J.W. Demmel, S.C. Eisenstat, J.R. Gilbert, X.S. Li, J.W., and H. Liu. A supernodal approach to sparse partial pivoting. *SIAM J. Matrix Analysis and Applications.*, 20(3):720–755, 1999.

[155] X.S. Li and J.W. Demmel. Superlu_dist: A scalable distributed-memory sparse direct solver for unsymmetric linear systems. *ACM Trans. Mathematical Software.*, 29(2):110–140, 2003.

[156] R.C. Whaley, A. Petitet, and J.J. Dongarra. Automated empirical optimization of software and the ATLAS project. *Parallel Computing*, 27(1–2):3–35, 2001. Also available as University of Tennessee LAPACK Working Note #147, UT-CS-00-448, 2000 (www.netlib.org/lapack/lawns/lawn147.ps).

[157] M. Naumov. Incomplete-lu and Cholesky preconditioned iterative methods using cusparse and cublas. Technical report, NVIDIA, Santa Clara, California, 2011.

[158] W.D. Henshaw. A primer for writing pde solvers with overture. Technical report, Lawrence Livermore National Laboratory, Livermore, California, 2011.

[159] T. Williams and C. Kelley. gnuplot 4.4: An interactive plotting program. Technical report, 2011.

[160] R. Burden and J.D. Faires. *Numerical Analysis*. Thomson, Belmont, California, eighth edition, 2005.

[161] E.D. Batista and J.E. Castillo. Mimetic schemes on non-uniform structured meshes. *Electronic transactions on numerical analysis*, 34:152–162, 1993.

[162] R.E. Ewing, editor. *The Mathematics of Reservoir Simulation*. The Society for Industrial and Applied Mathematics, Philadelphia, PA, first edition, 1983.

[163] J. Bear, C. Braester, and P.C. Menier. Effective and relative permeabilities of anisotropic porous media. *Transport in Porous Media 2*, pages 301–316, 1987.

[164] H.K. Vu. A mimetic workbench for reservoir simulations. Master's thesis, San Diego State University, San Diego, CA, 2006.

[165] T. Arbogat, C.N. Dawson, P.T. Keenan, and M.F. Wheeler. Enhanced cell-centered finite differences for elliptic equations. *SIAM Journal of Sci. Computing*, 18, 1997.

[166] T. Arvogast, M. Wheeler, and I. Yotov. Mixed finite elements for elliptic problems with tensor coefficients as cell-centered finite differences. *SIAM J. Numer. Anal.*, 34(2):828–852, 1997.

[167] Q. Zhou and J.T. Birkholzer. On scale and magnitude of pressure build-up induced by large-scale geologic storage of CO2. *Greenhouse Gases Science and Technology*, 1:11–20, 2011.

[168] W. Pentland. The carbon conundrum. *Forbes.com*, 2008.

[169] C.P. Paolini, A.J. Park, C. Binter, and J.E. Castillo. An investigation of the variation in the sweep and diffusion front displacement as a function of reservoir temperature and seepage velocity with implications in CO_2 sequestration. *Proceedings of the International Energy Conversion Engineering Conference*, 2011.

[170] Y.K. Kharaka, D.R. Cole, S.D. Hovorka, W.D. Gunter, K.G. Knauss, and B.M. Freifeld. Gas-water-rock interactions in Frio Formation following CO_2 injection: implications for the storage of greenhouse gases in sedimentary basins. *Geology*, 34(7):577–580, 2006.

[171] A. Taflove. *Computational Electrodynamics: the Finite Difference Time-Domain Method*. Artech House, Boston, Massachusetts, first edition, 1995.

[172] A. Yefet and P.G. Petropoulos. A staggered fourth-order accurate explicit finite difference scheme for the time-domain Maxwell's equations. *J. Comput. Phys.*, 168:286–315, 2001.

[173] J. Fang. *Time domain finite difference computation for Maxwell's equations*. PhD thesis, University of California, Berkeley, California, 1989.

[174] J. Hyman and M. Shashkov. Mimetic discretizations for Maxwell's equations and the equations of magnetic diffusion. *PIERS*, 32:89–121, 2001.

[175] O. Rojas. *Modeling of rupture propagation under different friction laws using high-order mimetic operators*. PhD thesis, San Diego State University and Claremont Graduate University, California, 2009.

[176] O. Rojas. Mimetic Finite Difference Modeling of 2D Elastic P-SV Wave Propagation. Qualifying exam report for the joint doctoral program in computational science, San Diego State University and Claremont Graduate University, San Diego, California, 2006.

[177] R. Madariaga. Dynamics of an expanding circular fault. *Bull. Seis. Soc. Am.*, 66:163–182, 1976.

[178] E. Saenger, N. Gold, , and S. Shapiro. Modeling the propagation of elastic waves using a modified finite-difference grid. *Wave Motion*, 31:77–92, 2000.

[179] A. Levander. Fourth-order finite-difference P-SV seismograms. *Geophysics*, 53:1425–1436, 1988.

[180] J. Kristek, P. Moczo, and R. Archuleta. Efficient methods to simulate planar free surface in the 3D 4th-order staggered-grid finite-difference schemes. *Stud. Geophys. Geod.*, 46:355–381, 2002.

[181] E. Gottschammer and K.B. Olsen. Accuracy of the explicit planar free-surface boundary condition implemented in a fourth-order staggered grid velocity-stress finite-difference scheme. *Bull. Seis. Soc. am.*, 91(3):617–623, 2001.

[182] J. Hyman, J. Morel, M. Shashkov, and S. Steinberg. Mimetic finite difference methods for diffusion equations. *Comput. Geosciences*, 6:333–352, 2002.

[183] J. Virieux. P-sv wave propagation in heterogeneous media: velocity-stress finite-difference method. *Geophysics*, 49:1933–1957, 1986.

[184] M. Shuo, R. Archuleta, and P. Liu. Hybrid modeling of plastic p-sv wave motion: A combined finite-element and staggered-grid finite-difference approach. *Bull. seis. Soc. Am.*, 94:1557–1563, 2004.

[185] J.E. Marsden and A.J. Tromba. *Vector Calculus*. W.H. Freeman and Company, San Francisco, California, first edition, 1976.

[186] J.S. Hesthavena and E.M. Rønquist, editors. *Spectral and High Order Methods for Partial Differential Equations*, volume 76 of *Lecture Notes in Computational Science and Engineering*. Springer, New York, 2011.

Index

9 780367 380434